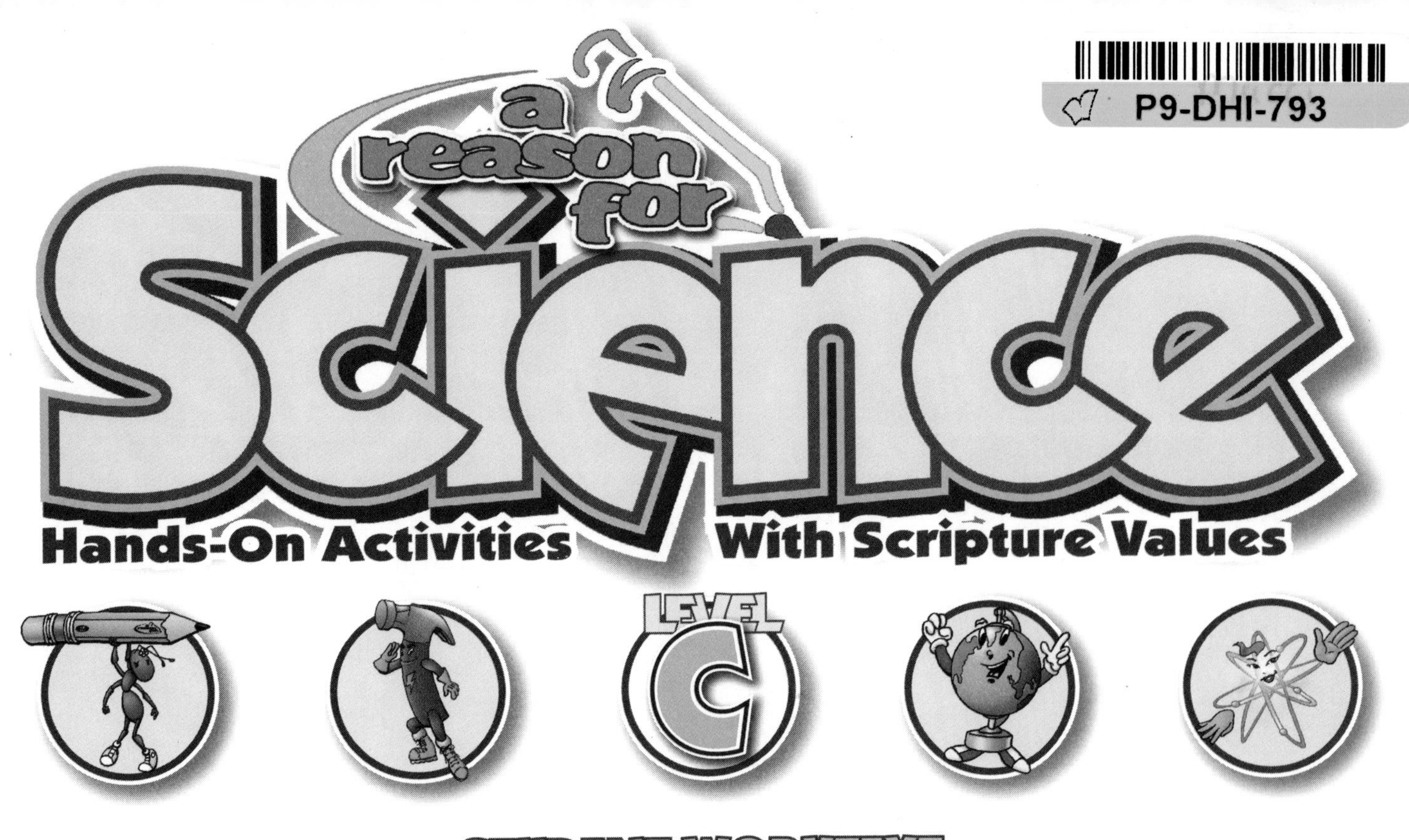

STUDENT WORKTEXT

ISBN #1-58938-142-4

Published by The Concerned Group, Inc.
700 East Granite • PO Box 1000 • Siloam Springs, AR 72761

Authors **Dave & Rozann Seela**
Publisher **Russ L. Potter, II**
Senior Editor **Bill Morelan**
Project Coordinator **Rocki Vanatta**
Creative Director **Daniel Potter**
Proofreader **Elizabeth Granderson**
Step Illustrations **Steven Butler**
Character Illustrations **Josh Ray**
Colorists **Josh & Aimee Ray**

Printed on recycled paper in the United States

For more information about **A Reason For®** curricula, write to the address above, call, or visit our website.

www.areasonfor.com
800.447.4332

Dear Parent,

Welcome to a new school year! This letter is to introduce you to **A Reason For©** Science.

A Reason For© Science teaches basic Life, Earth, and Physical Science through fun, hands-on activities. Each lesson is tied directly to the **National Science Education Content Standards** and uses an inquiry-based approach designed to enhance learning.

Today's increasingly complex world requires a clear understanding of science and technology. Our future prosperity depends on helping children rediscover the challenges, excitement, and joy of science — especially in the context of Scripture values. Thus, one of the primary goals of **A Reason For©** Science is to make science not only meaningful, but also FUN!

Fun, Flexible Format

Instead of a hardback textbook filled with "facts" to memorize, your child will be working in an interactive worktext designed to develop critical-thinking skills. Students start each week with a hands-on activity demonstrating a key science concept. This is followed by group discussion, journaling, and a series of thought-provoking questions. Lessons conclude with a summary of key concepts and a related "object lesson" from Scripture.

Safety Issues

The hands-on nature of **A Reason For©** Science means your child will be working with age-appropriate materials. (For instance, the "acids" we use are actually dilute forms comparable to typical household chemicals.) Like a field trip or gym class, these science activities usually require simple safety precautions.

But for instructional reasons, all materials in **A Reason For©** Science are treated as hazardous! This encourages students to develop good safety habits for use in later years. (If you have further questions about safety, your child's teacher has an in-depth safety manual outlining precautions for every lesson.)

Scripture Values

Best of all, **A Reason For©** Science features Scripture values! Every lesson concludes with a Scripture object lesson related to the week's topic. These "Food for Thought" sections encourage students to relate everyday experiences to Scriptural themes, providing a positive way to integrate faith and learning.

Here's to an exciting year exploring God's world!

Dave & Rozann Seela
Authors, **A Reason For©** Science

A Reason For© Science makes science FUN! Your school year will be filled with hands-on activities, colorful discovery sheets, and lots of discussion and exploration. You'll discover many exciting new things as you explore God's world!

Although "almost everything relates to everything else" in some way or another, scientists usually divide science into three broad areas for study: **life**, **earth**, and **physical science**. The sections in your worktext are based on these categories.

Colorful icons are used to help you identify each section. An **ant** represents **Life Science** lessons. A **globe** stands for **Earth Science** lessons. An **atom** introduces **Physical Science (Energy/Matter)** lessons. And a **hammer** represents the **Physical Science (Forces)** lessons.

Life Science

Life Science is the study of **living things.** In the Life Science section of your **A Reason For©** Science worktext, you'll explore different kinds of living things. You'll learn about their characteristics (how they're alike or different). You'll discover how scientists classify (label and sort) living things. You'll even learn more about your own body and how it works!

Earth Science

Earth Science is the study of **earth** and **sky**. In the Earth Science section of your **A Reason For©** Science worktext, you'll explore the structure of our planet (rocks, crystals, volcanos), the atmosphere (air, clouds), and related systems (water cycles, air pressure, weather). Grades 7 and 8 reach out even further with a look at the solar system and stars.

Physical Science

Physical Science is the study of **energy, matter,** and related **forces.** Since the physical part of science has a big effect on your daily life, it's divided into two major sections:

Energy and Matter

In this section of your **A Reason For©** Science worktext, you'll learn about the different **states** and unique **properties** of **matter**. You'll discover new things about light and sound. You'll explore physical and chemical reactions. Some grades will explore related concepts like circuits, currents, and convection.

Forces

In this section of your **A Reason For©** Science worktext, you'll discover how "**push** and **pull**" form the basis for all physical movement. You'll explore simple machines (levers, pulleys). You'll work with Newton's laws of motion. You'll even learn to understand concepts like torque, inertia, and buoyancy.

Safety First!

Before you begin, be sure to read about "Peat" the safety worm. Peat's job is to warn you whenever there's a potential hazard around. (Whenever you see Peat and his warning sign, STOP and wait for further instructions from your teacher!)

LEARN TO BE SAFE!

Exploring God's world of science often requires using equipment or materials that can injure you if they're not handled correctly. Also, many accidents occur when people hurry, are careless, or ignore safety rules.

It's your responsibility to know and observe the rules and to use care and caution as you work. Just like when you're on the playground, horseplay or ignoring safety rules can be dangerous. Don't let an accident happen to you!

Meet "Peat" the Safety Worm!

Peat's job is to warn you whenever there's a potential hazard around. Whenever you see Peat and his warning sign, **STOP** and wait for further instructions from your teacher!

Peat's sign helps you know what kind of hazard is present. Before beginning each activity, your teacher will discuss this hazard in detail and review the safety rules that apply.

means this activity requires **PROTECTIVE GEAR.**

Usually **gloves** or **goggles** (or both) are required. Goggles protect your eyes from things like flying debris or splashing liquids. Gloves protect your hands from things like heat, broken glass, or corrosive chemicals.

means there is a **BURN HAZARD.** There are three common burn hazards.

"Open Flame" indicates the presence of fire (often matches or a candle). **"Thermal Burn"** means objects may be too hot to touch. **"Corrosion"** indicates a chemical substance is present.

LEARN TO BE SAFE!

CONTINUED

means there is a POISON HAZARD.

There are three common poison hazards. "Skin Contact" indicates a substance that should not touch skin. "Vapor" indicates fumes that should not be inhaled. "Hygiene" indicates the presence of materials that may contain germs.

indicates **OTHER HAZARDS.**

There are three additional hazards that require caution. **"Breakage"** indicates the presence of fragile substances (like glass). **"Slipping"** indicates liquids that might spill on the floor. **"Sharp Objects"** indicates the presence of tools with sharp edges or points.

Play It Safe!

Exploring God's world with A Reason For© Science can be great fun, but remember — play it safe! Observe all the safety rules, handle equipment and materials carefully, and always be cautious and alert.

And don't forget Peat the safety worm! Whenever you see Peat and his warning sign, **STOP** and wait for further instructions from your teacher.

Life
CHAPTERS 1-9
FRIGI-BREEZE

Life Science

Life Science is the study of **living things.** In this section, you'll explore different kinds of living things. You'll learn about their characteristics (how they're alike or different). You'll discover how scientists classify (label and sort) living things. You'll even learn more about your own body and how it works!

Week

NAME ______________________

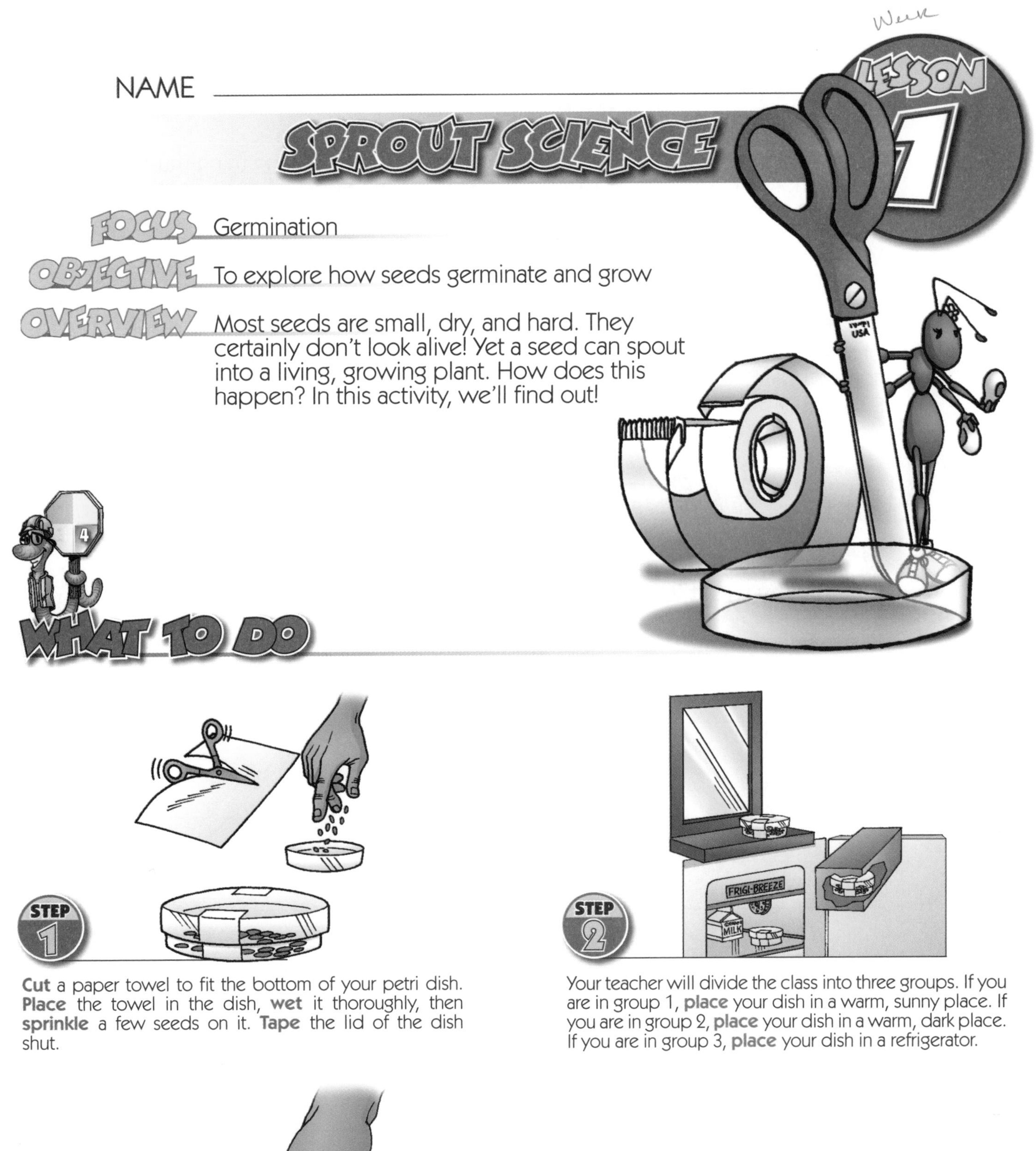

SPROUT SCIENCE

FOCUS Germination

OBJECTIVE To explore how seeds germinate and grow

OVERVIEW Most seeds are small, dry, and hard. They certainly don't look alive! Yet a seed can spout into a living, growing plant. How does this happen? In this activity, we'll find out!

WHAT TO DO

STEP 1

Cut a paper towel to fit the bottom of your petri dish. **Place** the towel in the dish, **wet** it thoroughly, then **sprinkle** a few seeds on it. **Tape** the lid of the dish shut.

STEP 2

Your teacher will divide the class into three groups. If you are in group 1, **place** your dish in a warm, sunny place. If you are in group 2, **place** your dish in a warm, dark place. If you are in group 3, **place** your dish in a refrigerator.

STEP 3

Examine your dish every day and **draw** pictures of what you see. If the paper towel looks like it's getting dry, **open** the dish and **add** a little more water. Be sure to **reseal** the lid with tape.

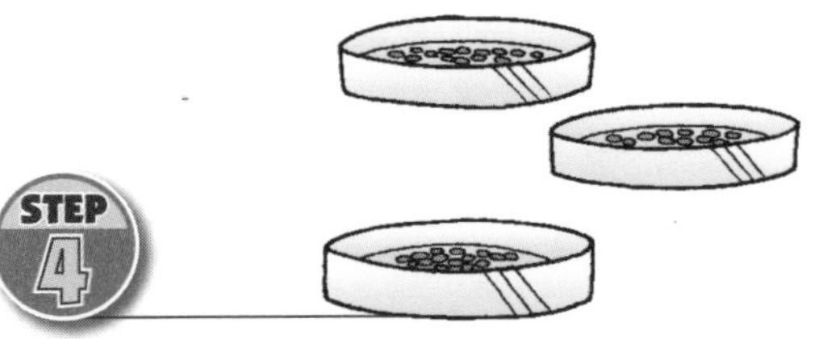

STEP 4

After a few days, **compare** your dish with the dishes of others in your group. Now **compare** your dish with the dishes of the other two groups. **Discuss** what you have observed with your research team.

Nothing on earth lives forever. For life to continue, every living thing must **reproduce** (make more of itself). Most plants reproduce using **seeds**. Seeds contain two basic parts: an **embryo** (baby plant) and an **energy source** (food).

When seeds **sprout** (begin to grow), scientists call it **germination**. As we saw in this activity, germination requires the correct **environment**.

American farmers produce huge amounts of food. Since many crops begin as seeds, proper germination is very important. Farmers use expensive machinery to prepare the soil and plant seeds just so. Then, as long as the weather cooperates, the seeds will grow — producing plenty of food to eat!

Why was the paper towel in the bottom of the dish important? Why was it important to tape the lid shut?

In Step 2, what things were the same between groups?

In Step 2, what things were different between groups?

When the activity was completed, how were the results different between groups?

Based on what you've learned, what are some things that plants need in order to germinate?

Seeds contain an embryo and food for the embryo to use until it grows. To germinate, seeds need proper amounts of warmth, moisture, and sunlight. If any one of these is missing, the growing plant may not survive.

Genesis 1:11 Sometimes we forget that signs of God's love are everywhere. Scripture tells us God blessed the Earth with fruitful plants and good things to eat. If you're feeling sad or lonely, think about these blessings.

Sprouting seeds and growing plants are miracles that happen every day — providing a constant reminder of how much God loves you!

Week

NAME ____________________

SATURATION SITUATION

LESSON 2

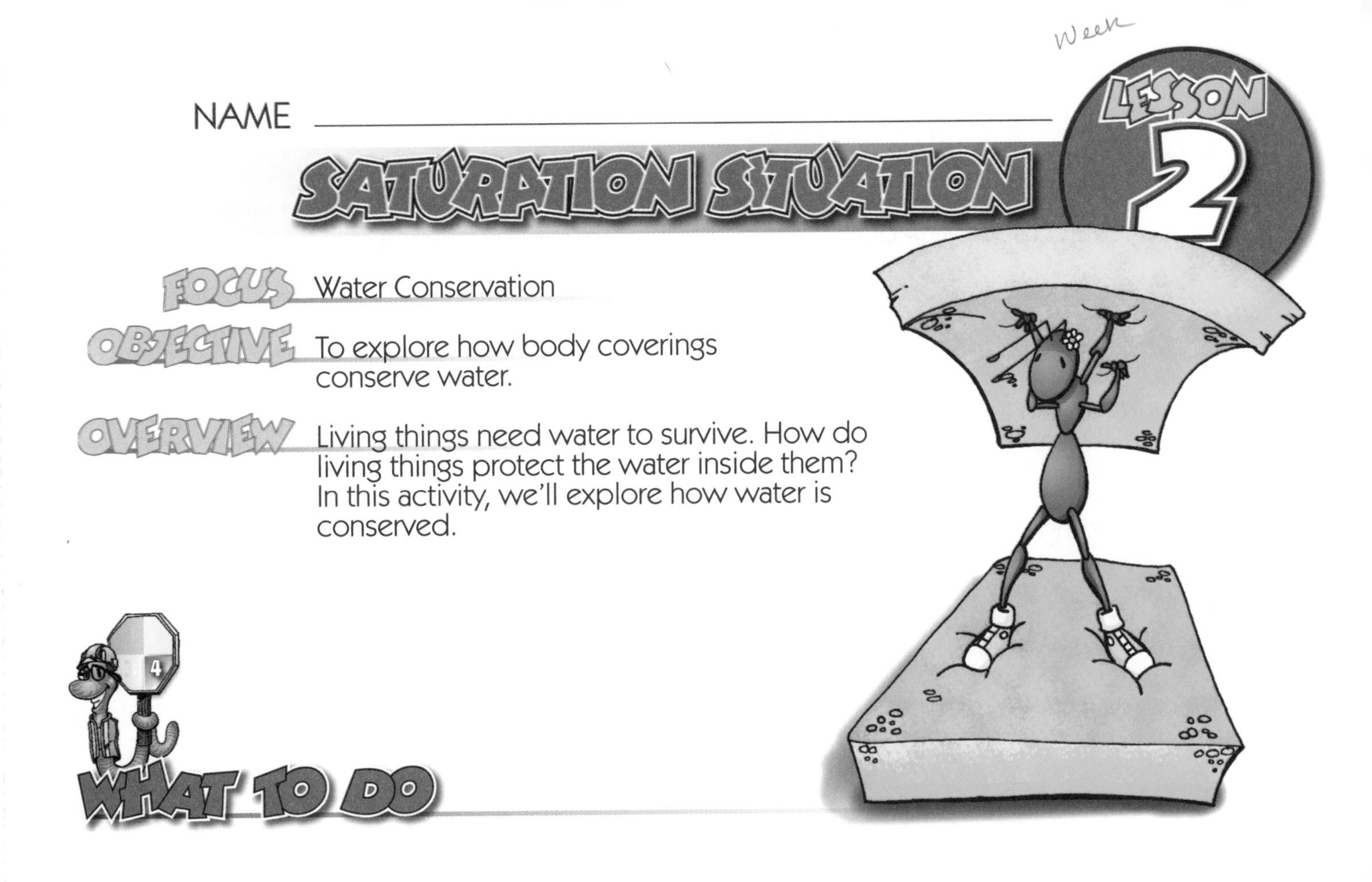

FOCUS Water Conservation

OBJECTIVE To explore how body coverings conserve water.

OVERVIEW Living things need water to survive. How do living things protect the water inside them? In this activity, we'll explore how water is conserved.

WHAT TO DO

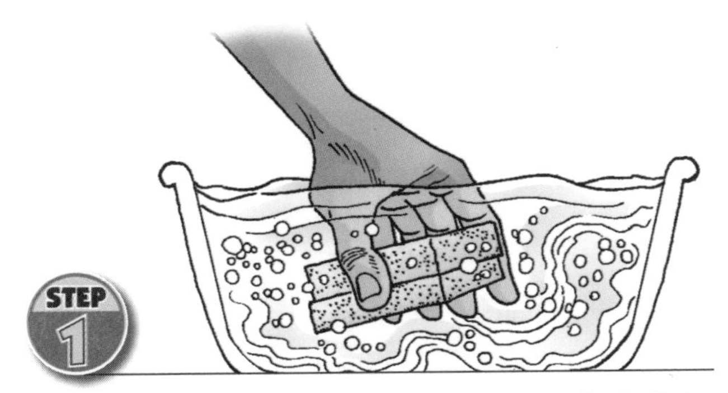

STEP 1

Dip two sponges in water until they are soaked all the way through. Make sure they are soaked equally.

STEP 2

Completely **wrap** one sponge in plastic wrap. Carefully **check** to make sure there are no openings anywhere in the plastic.

STEP 3

Lay the sponges side by side in a pie pan or tray. **Place** the pan in a warm, sunny place. **Predict** what you think will happen to the two sponges.

STEP 4

Wait two days, then **check** your sponges to see what's happened. **Record** the results in your journal. **Review** each step in this activity. **Discuss** what you've observed with your research team.

What Happened?

The plastic wrap **modeled** one function of your **skin** — conserving water. The covered sponge kept its water. The sponge with no covering dried out.

Although you used only one **layer** of material for this activity, human skin has several layers. Each of these layers has its own unique **structure** (how it's built) and **function** (purpose). Together they provide many different kinds of protection.

For instance, to protect against damage from sunlight, your skin produces a chemical called **melanin**. The amount of melanin in your skin determines your skin color. Darker skin has higher levels of melanin, providing more protection. Lighter skin has lower levels of melanin. When exposed to sunlight, lighter skin produces additional melatin, creating a "tan." As you can see, skin is a complex and amazing protective device.

What does the sponge represent in this activity?
What does the water represent?

What does the plastic wrap represent?
Why was it important to make sure there were no openings?

3. What did you predict in Step 3?
How did this prediction reflect what actually happened?

Skin color varies not only from person to person, but even on the same person over time. What causes this?

5. Based on what you've learned, explain why body coverings are important. What might happen to a living thing that lost its body covering?

Conclusion

Living things need water to survive. A body covering (like skin) helps living things conserve water.

Food for Thought

Psalm 40:11, 12 This activity demonstrates one amazing ability of human skin — conservation of water. Without the protective covering of our skin, we'd dry out like an uncovered sponge! Skin helps keep essential water inside, protecting us from harm.

Sometimes people think they can make it entirely on their own. They use the gifts God has given them for their own personal pleasure. But without the protective covering of God's love and faithfulness, their hearts soon become barren and dry.

Wrap yourself in God's love — the only real protection!

Journal My Science Notes

NAME ______________________________

LESSON 3

TUBE EYE

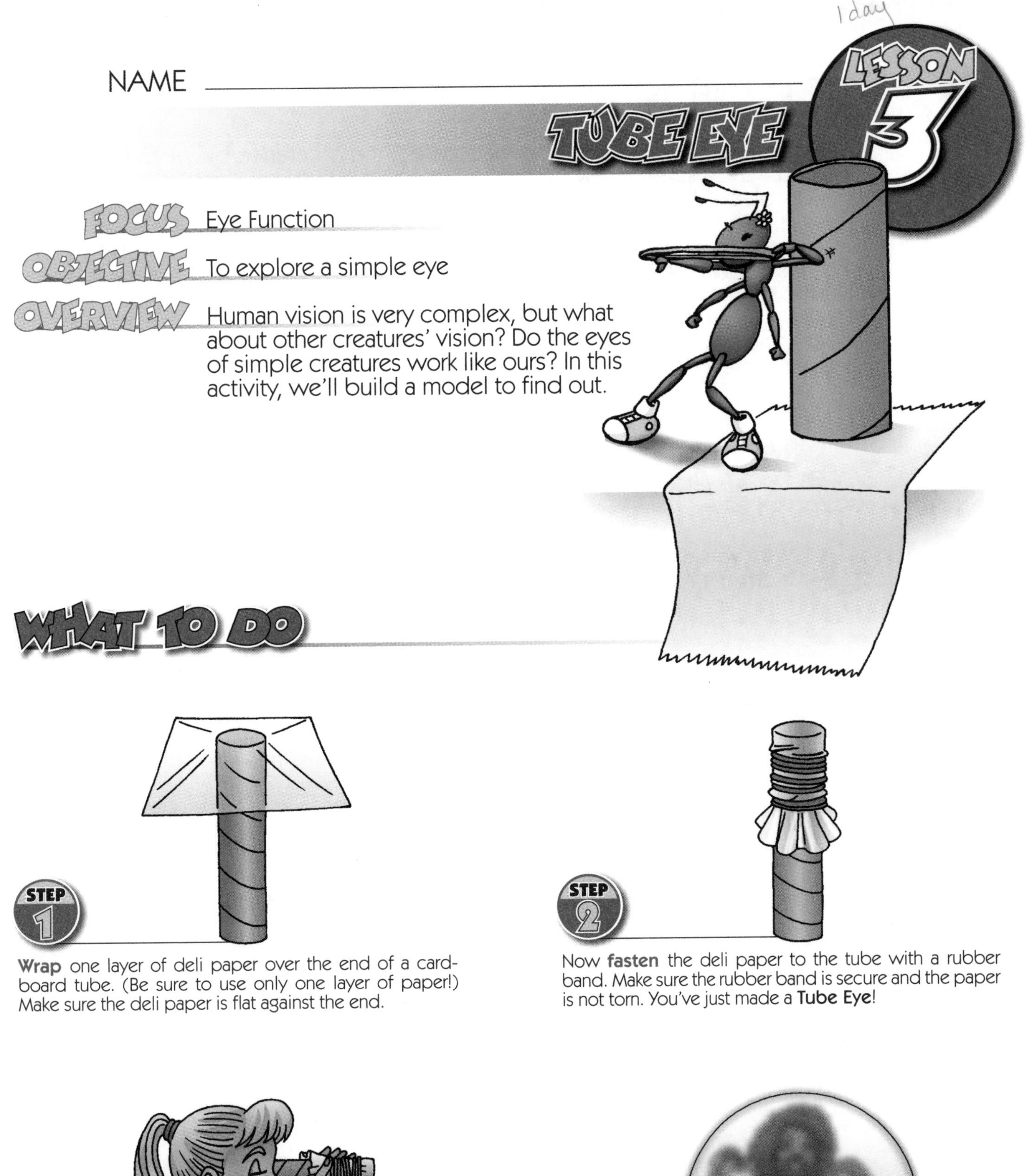

FOCUS Eye Function

OBJECTIVE To explore a simple eye

OVERVIEW Human vision is very complex, but what about other creatures' vision? Do the eyes of simple creatures work like ours? In this activity, we'll build a model to find out.

WHAT TO DO

STEP 1

Wrap one layer of deli paper over the end of a cardboard tube. (Be sure to use only one layer of paper!) Make sure the deli paper is flat against the end.

STEP 2

Now **fasten** the deli paper to the tube with a rubber band. Make sure the rubber band is secure and the paper is not torn. You've just made a **Tube Eye**!

STEP 3

Close one eye (or cover it with your hand). **Hold** the open end of your **Tube Eye** up to your other eye and **look** through it. **Experiment** by looking toward a bright light, then looking toward a dark wall.

STEP 4

Repeat Step 3, but this time **sit** about three feet away from your research team and **look** toward them. **Describe** what you see. Now **review** each step in this activity. **Discuss** what you've observed with your research team.

What Happened?

The simple eye you created could only detect the difference between **light** and **dark** and maybe sense a few rough shapes. Since simple creatures usually feed in very dark places, this type of **vision** is really all they need.

Human vision is much more complex. The human eye uses a **lens** to bend (**focus**) light. This gathers the light together, allowing us to see detail (not just shapes), and gives the **brain** the information it needs to understand complex **images**.

Just imagine how restricted your life would be if you could only see the world through simple eyes!

Why was it important to use only one layer of deli paper in Step 1?

How is the device you built similar to a telescope? How is it different?

Describe what you saw when you looked through the Tube Eye in Step 3.

4 Describe what you saw in Step 4. What difference did it make if light was behind or in front of your team members?

5 How is human vision different from the vision of simple creatures? What makes our eyes more complex?

CONCLUSION

All kinds of eyes gather light to help living things make sense of the world around them. God gave different creatures different kinds of eyes to meet their specific needs.

Proverbs 3:5, 6 Imagine spending an entire school day using only a **Tube Eye** for vision! You'd really have to trust your friends and depend on them to help you make it through the day.

God sees and knows so much more than we do. By comparison we're almost blind! But this Scripture reminds us that we can always depend on God's guidance. Learn to trust God completely. Put God first in everything you do, and you can walk through this world with confidence!

My Science Notes

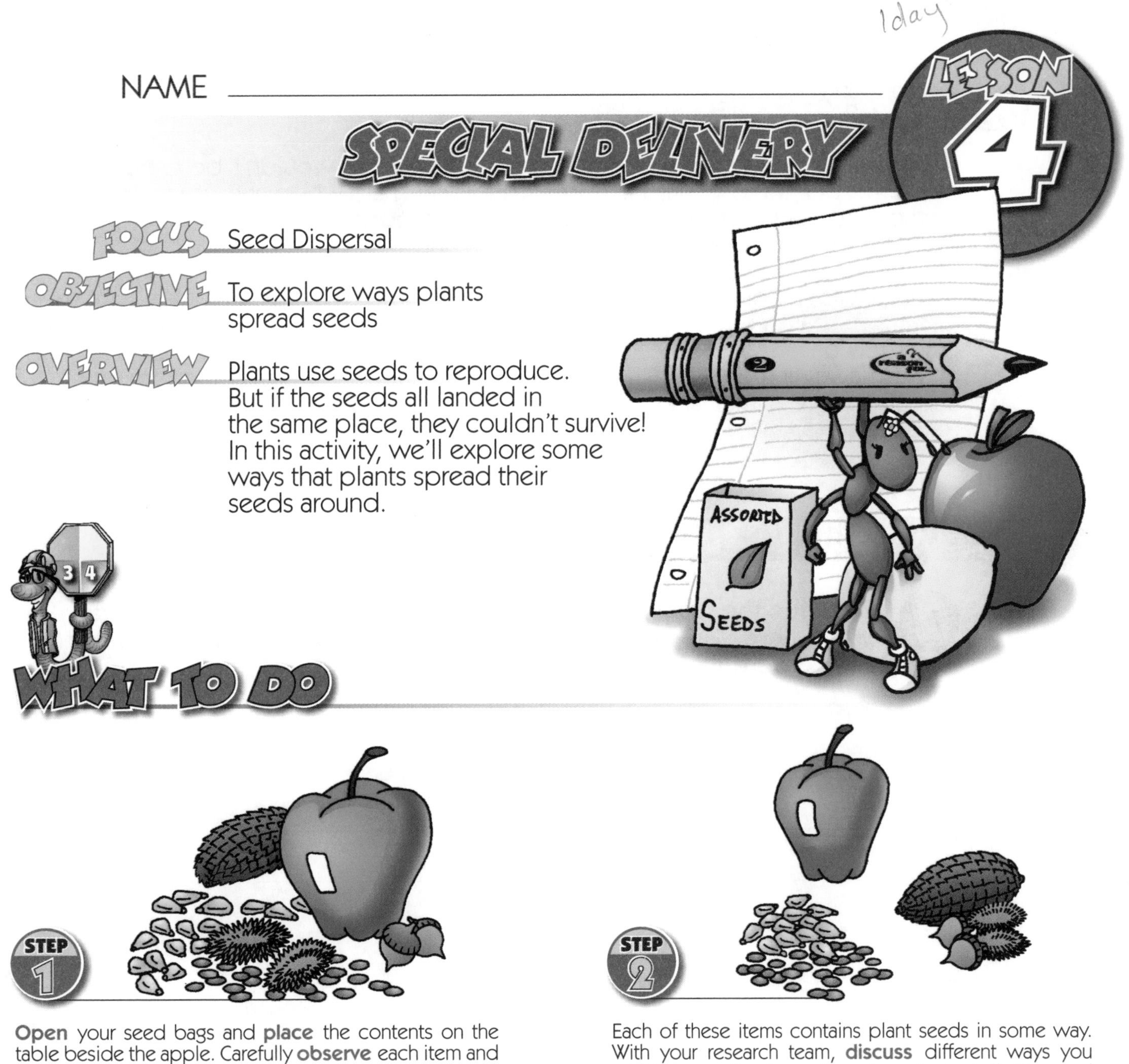

NAME ____

LESSON 4

SPECIAL DELIVERY

FOCUS Seed Dispersal

OBJECTIVE To explore ways plants spread seeds

OVERVIEW Plants use seeds to reproduce. But if the seeds all landed in the same place, they couldn't survive! In this activity, we'll explore some ways that plants spread their seeds around.

WHAT TO DO

STEP 1

Open your seed bags and **place** the contents on the table beside the apple. Carefully **observe** each item and **make notes** about what you see.

STEP 2

Each of these items contains plant seeds in some way. With your research team, **discuss** different ways you might group the items based on their similarities and differences.

STEP 3

Watch as your teacher cuts your team's apple in half. **Examine** the inside of the apple closely, then **find** and **remove** the seeds. **Discuss** what you see with your research team. (Remember, good scientists don't EAT their experiments!)

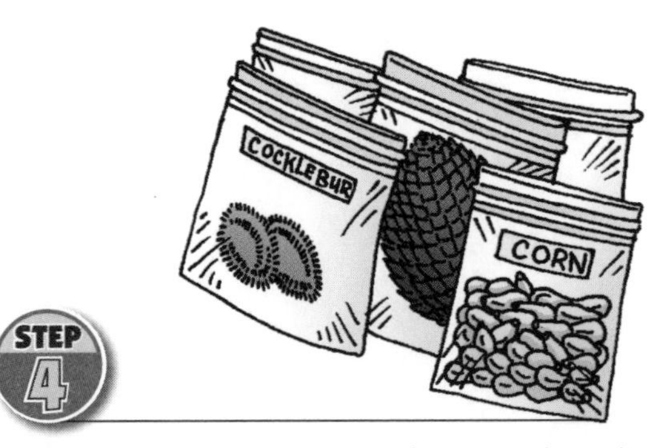

STEP 4

After making notes, **dispose** of your apple as the teacher directs. **Place** all the other seeds back into the container. Based on what you have observed, **predict** how each of the "seed packages" you've examined gets "delivered." **Discuss** your ideas with your research team.

If a plant's **seeds** all landed in the same place, there wouldn't be enough **water**, **nutrients**, or **light** to go around. Most would die! That's why plants use many different methods to spread their seeds. This gives the **embryo** (baby plant) in each seed a much better chance of becoming a mature plant.

Some plants rely on **wind** to spread their seeds. They may produce very light seeds (ash tree), "parachute" seeds (milkweed), or even seeds with wings (maple). Some plants rely on **water** to spread their seeds. They produce seeds that float (coconuts) or wash away in heavy rains (grasses). Some plants even rely on the **movement** of animals to spread their seeds. They may produce edible **fruit** (berries), or tiny hooks (cocklebur) to grab a ride.

Describe the items from the seed bags in Step 1. How many kinds were there? How many of each? How were they similar? How were they different?

Describe the groups your team made in Step 2. What characteristics did you use to sort the seeds?

What is the baby plant inside a seed called?
Name three things that it needs to survive.

List three different methods plants use to disperse their seeds.
Give an example of each.

A plant produces many more seeds than it needs to replace itself.
Based on what you've learned, why is this necessary?

To ensure survival, plants must spread seeds over great distances. Seed dispersal methods include wind, water, and the movement of animals.

Matthew 13:1-23 This parable describes some of the hardships seeds face. Some are eaten up, some sprout in poor soil and don't live long, some are choked out by other plants, and a few fall on good soil where they grow and prosper.

Jesus told parables about common things to help his listeners understand spiritual things. This parable reminds us that hearing God's word is not enough. For God's love to grow and prosper in our hearts, we must let Jesus in to prepare the way. The more time we spend with Jesus, the more open our hearts become to the power of God's presence.

1 day

NAME

LESSON 5

LEAF UMBRELLA

FOCUS Plant Structure

OBJECTIVE To explore how leaves direct rain

OVERVIEW Almost all plants have leaves. Without leaves, a plant would die. Leaves serve many different functions. In this activity, we'll explore how leaves can control the direction of raindrops to help and protect the plant.

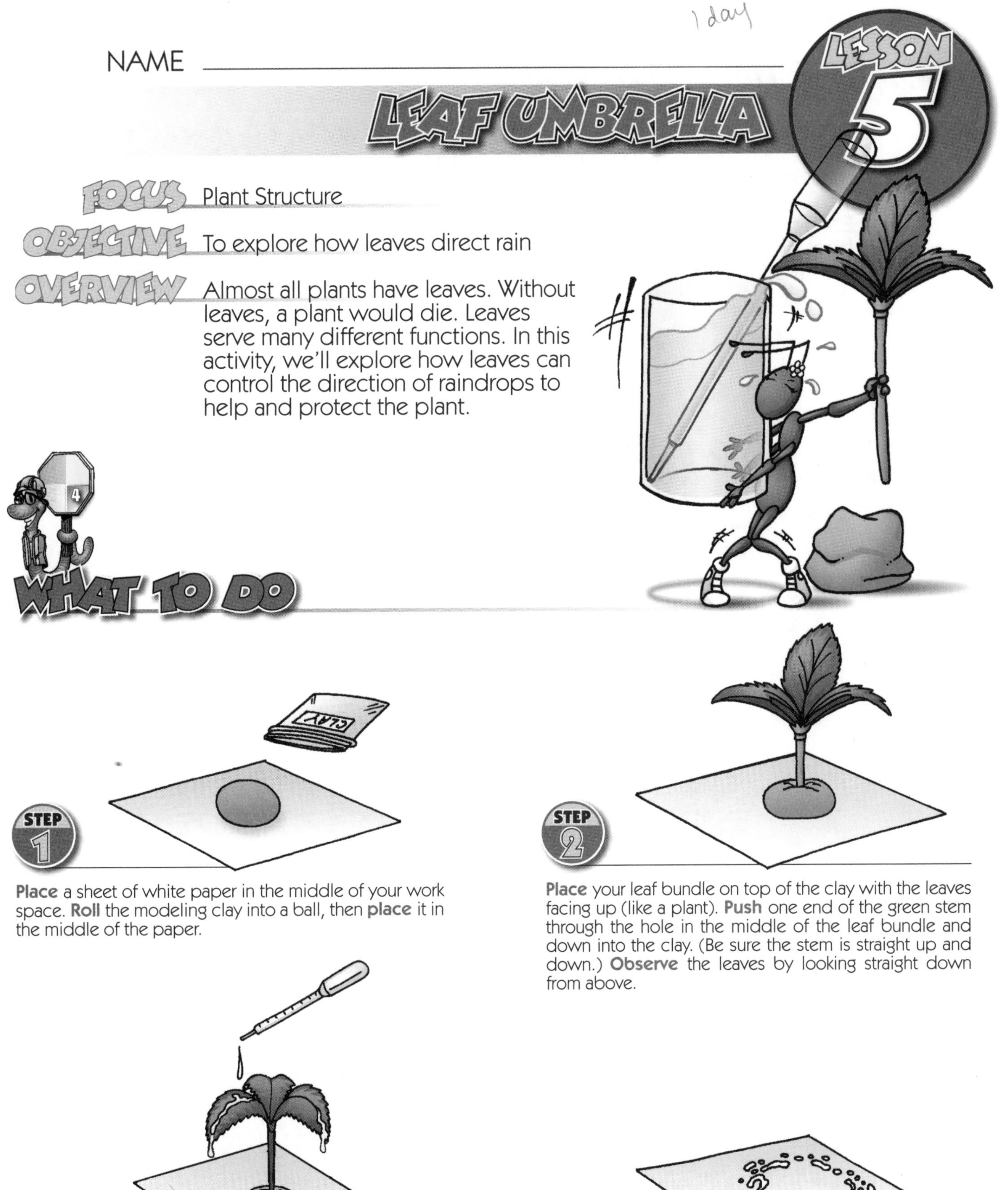

WHAT TO DO

STEP 1

Place a sheet of white paper in the middle of your work space. **Roll** the modeling clay into a ball, then **place** it in the middle of the paper.

STEP 2

Place your leaf bundle on top of the clay with the leaves facing up (like a plant). **Push** one end of the green stem through the hole in the middle of the leaf bundle and down into the clay. (Be sure the stem is straight up and down.) **Observe** the leaves by looking straight down from above.

STEP 3

Hold your pipette a few inches above the leaves. **Release** a drop of water. When the drop hits a leaf, **watch** where it ends up. **Move** your pipette around and **drop** water from different places. **Drip** and **drop** and **observe**, but don't soak the white paper too much.

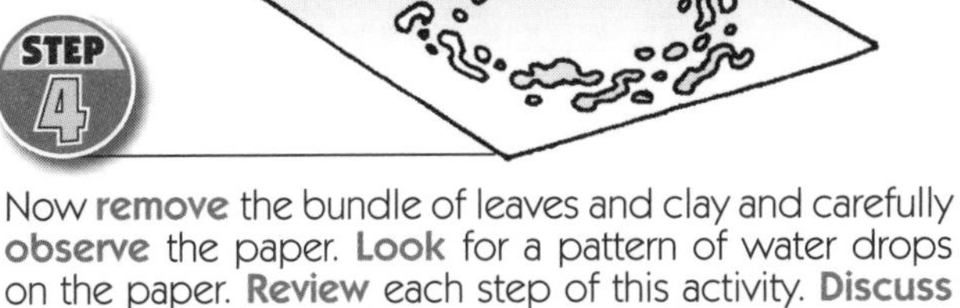

STEP 4

Now **remove** the bundle of leaves and clay and carefully **observe** the paper. **Look** for a pattern of water drops on the paper. **Review** each step of this activity. **Discuss** what you've observed with your research team.

What Happened?

Leaves serve many different functions. In this activity, we saw how a **plant's** leaves help control the direction of raindrops. The leaves spread the falling water evenly to the **roots**, helping them **absorb** water more efficiently.

Each raindrop lost some of its **kinetic** (moving) **energy** when it hit a leaf. The leaf bent or bounced slightly with each drop, absorbing the energy before the drop hit the ground. This protective action helps keep the **soil** around the plant from **eroding** (washing away) during heavy rains.

What We Learned

1. Why was it important to spread the leaves evenly in Step 1? What did the ball of clay represent?

2. How is a plant helped by its leaves' ability to control the direction of raindrops? What part of the plant benefits most?

3 Describe how leaves can help keep the soil around a plant from washing away.

4 Describe the pattern left on the paper in Step 4.
How does this pattern reflect the work of the leaves?

Based on what you've learned, what kinds of things could happen to a plant that only had leaves on one side?

Leaves serve many different functions. Two of these include directing rain to the plant's roots and absorbing the energy of falling water to decrease the chance of erosion.

Psalm 61:4 In this activity, we learned how leaves contribute to protecting a plant. When the rain begins to fall, leaves absorb the energy of the falling water. They also divert the water to where it can do the most good.

Just as leaves help protect a plant, so God's love protects his children. This Scripture reminds us that no matter what is happening around us, when we put our trust in God, we can be kept safe from those who would harm us.

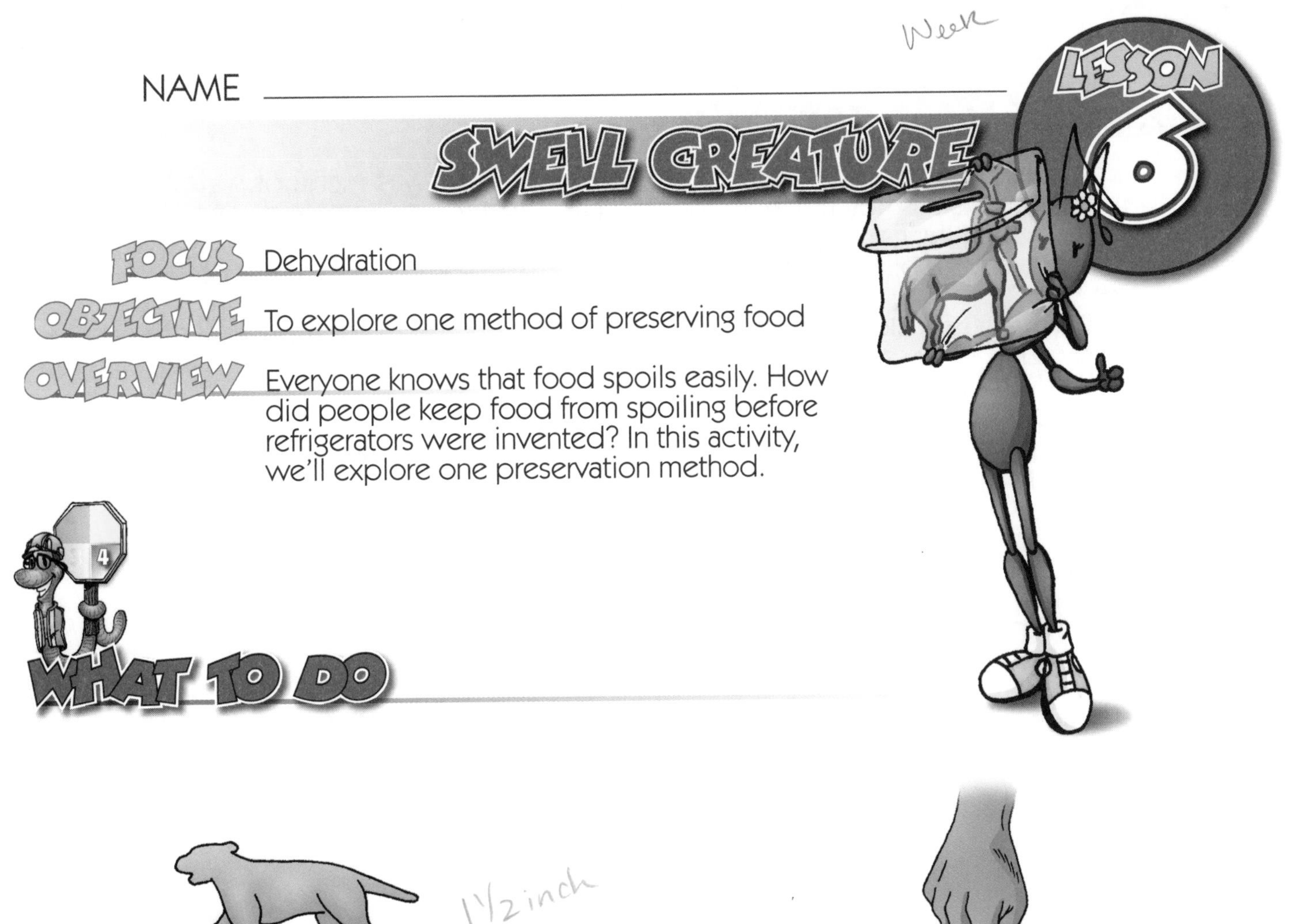

NAME ______________________________

LESSON 6

SWELL CREATURE

FOCUS Dehydration

OBJECTIVE To explore one method of preserving food

OVERVIEW Everyone knows that food spoils easily. How did people keep food from spoiling before refrigerators were invented? In this activity, we'll explore one preservation method.

WHAT TO DO

STEP 1

Carefully **observe** your "swell creature." Use a ruler to **measure** the creature. How long is it? How tall is it? How thick is it? **Record** these measurements.

STEP 2

Fill your bowl with clean water. Gently **place** your creature into the water. **Describe** what the creature looks like in the water, then **set** the container aside.

STEP 3

[next day] **Describe** what your creature looks like now! Carefully **lift** it from the water and let the excess water drip back in the bowl. After a few moments, **repeat** the measurements from Step 1 and **record** the changes. **Discuss** what you've observed with your research team.

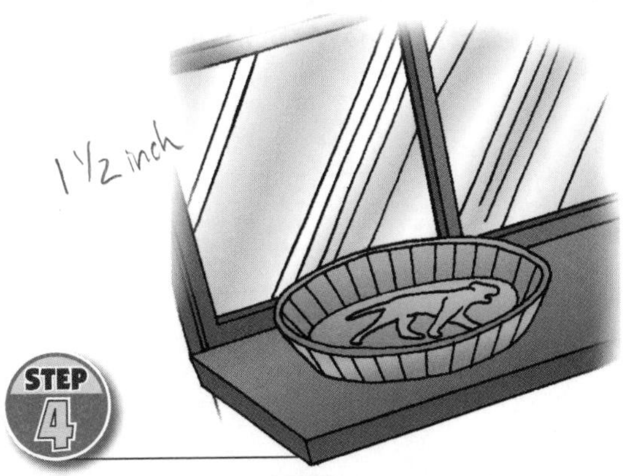

STEP 4

Place your creature in a pie pan or tray in a sunny window. Let it dry until the end of the week. Now **measure** it again and **record** the results. Based on your observations, **predict** what might happen if it dried several more days. **Discuss** your observations and predictions with your research team.

Just like food, a large part of your **Swell Creature** was made of **water**. You could tell because it got much smaller after the water was removed. **Micro-organisms** (tiny living things) like **bacteria** and **mold** need warmth and water to grow. Without adequate supplies of both, food **decomposes** (spoils) much more slowly.

Refrigeration slows spoilage by removing warmth. **Dehydration** slows the process by removing essential water. Since earliest times, people have used dehydration to help preserve food for later use.

Some forms of food made by dehydration (beef jerky, raisins, dried fruit) are still very popular today!

Compare the Swell Creature in Step 1 with the Swell Creature in Step 3. How were they similar? How were they different?

What did you predict in Step 4?
How did this prediction reflect what actually happened?

Name two things bacteria or mold needs to grow.
When micro-organisms thrive, what happens to food?

4 Why does dehydration help preserve food?
Give two examples of dehydrated foods.

5 What are some other methods people use to preserve food?

Without preservation, food spoils rapidly. One way to preserve food is to remove most of the water. This process is called dehydration.

John 7:38 Your little creature was dry and empty, and it looked really small. But when you placed it in the bowl, it began to absorb a lot of water. It wasn't long before it looked like a new creature!

Sometimes our hearts are dry and empty. We feel very small inside. But this Scripture reminds us that we can be filled with "living water" that comes from Jesus! As we let God's goodness work in our lives, we become new creatures, filled with God's love.

My Science Notes

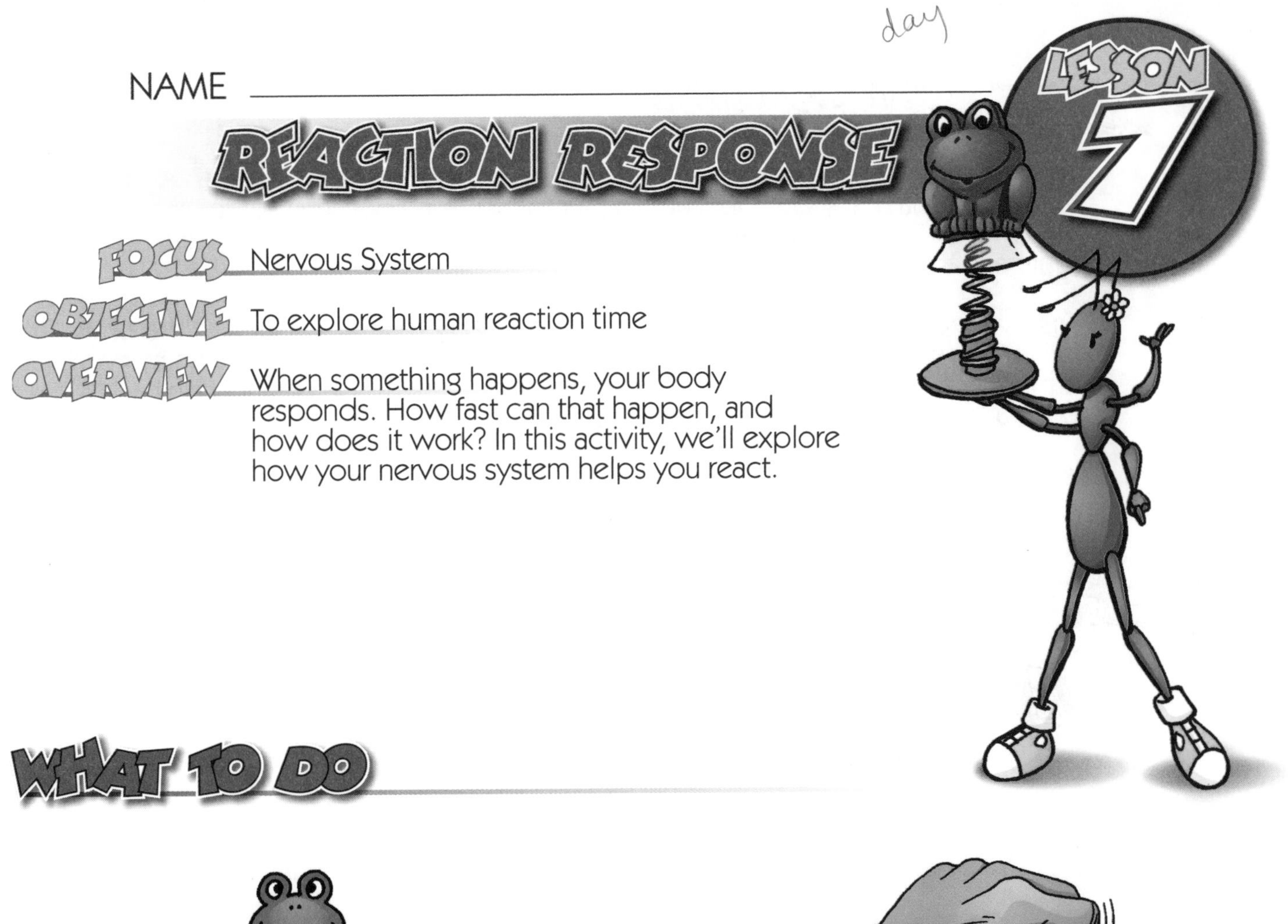

NAME ______________________________

REACTION RESPONSE

LESSON 7

FOCUS Nervous System

OBJECTIVE To explore human reaction time

OVERVIEW When something happens, your body responds. How fast can that happen, and how does it work? In this activity, we'll explore how your nervous system helps you react.

WHAT TO DO

STEP 1

Carefully **examine** your frog. **List** the four major parts of the frog device.

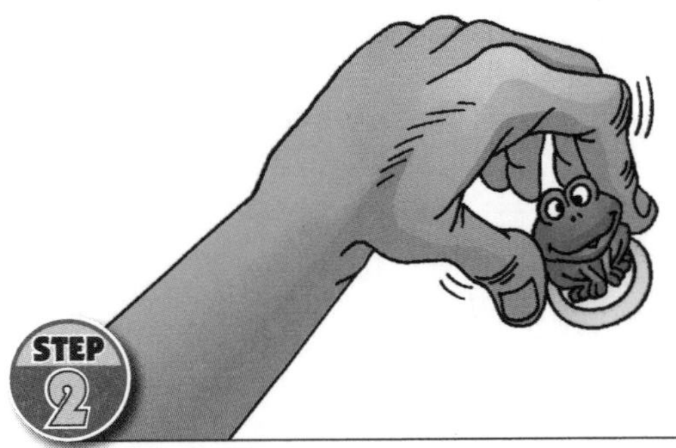

STEP 2

Place the frog on your work surface and **push down** until the suction cup sticks to the stand. **Observe** the frog and **watch** what happens. Be very patient! After something happens, **discuss** what you've observed and why you think it happened.

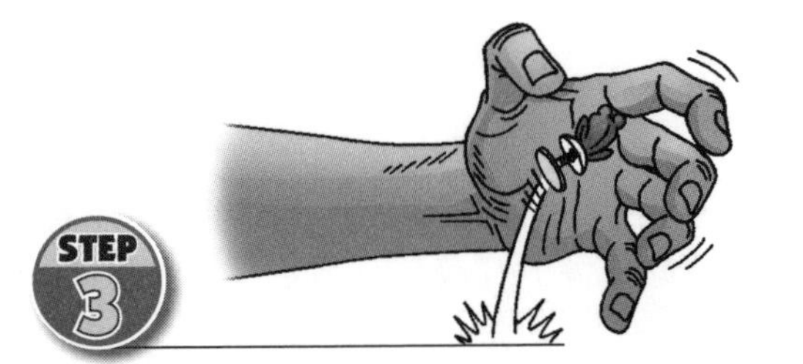

STEP 3

Retrieve the frog. **Stand** facing the frog with your hands behind your back. **Ask** a team member to "reset" the frog. Now **watch** closely. When the frog jumps, try to **grab** it before it hits the table. **Repeat** this three times. After everyone on your research team has had a turn, **record** the results.

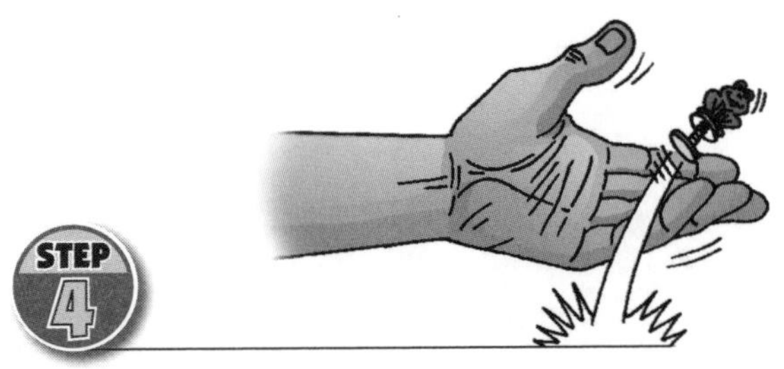

STEP 4

Reset the frog again. **Stand** with your back to the frog, hands at your sides. **Ask** a team member to watch the frog. When it jumps, they must say, "Now!" Quickly **turn** and try to **grab** the frog before it hits the table. After everyone has had a turn, **record** the results. **Review** each step in this activity. **Discuss** your observations with your research team.

Scientists call the process you just observed **stimulus/response**. A stimulus is something that happens (like the frog jumping). A response is something your body does when the stimulus happens (like trying to catch the frog). Although we don't realize it, this can be a very complicated process!

First, your **eyes** had to see the frog jump and **relay** that information to the **brain**. The brain then sent a signal ("Catch that frog!") through your **nerves** to the **muscles** in your arm and hand. Since both the frog and your arm were moving, your brain had to make constant, tiny adjustments to compensate.

As you can see, God designed your marvelous nervous system to perform complex movements rapidly and smoothly.

What is a stimulus? What is a response? Give an example of a stimulus/response.

Compare Step 3 with Step 4. How were they similar? How were they different? What additional sense was used in Step 4?

Describe the difference in reaction times between your team mates. Why might God give different people different kinds of skills?

4 Give another example of a stimulus/response. List the main body parts and senses involved in reacting to this stimulus.

5 Alcohol slows stimulus/response times. Based on what you've learned, explain why even one drink can make someone a more dangerous driver.

A stimulus is something that happens. A response is your reaction. For the stimulus/response process to work properly, several parts of your body must work together.

Ephesians 4:16 In this lesson, you discovered that when something happens (a stimulus), your body reacts (a response). This is an important process in many games. To achieve the goal you have in mind, all the parts of your body must work together!

This Scripture tells us that a church functions like that. All the parts must work together for the good of all. We not only need someone to conduct the services, but also to sweep the floor, mow the grounds, and maintain the buildings. Take time this week to thank the people who make your church run so well!

My Science Notes

day

NAME ______________________________

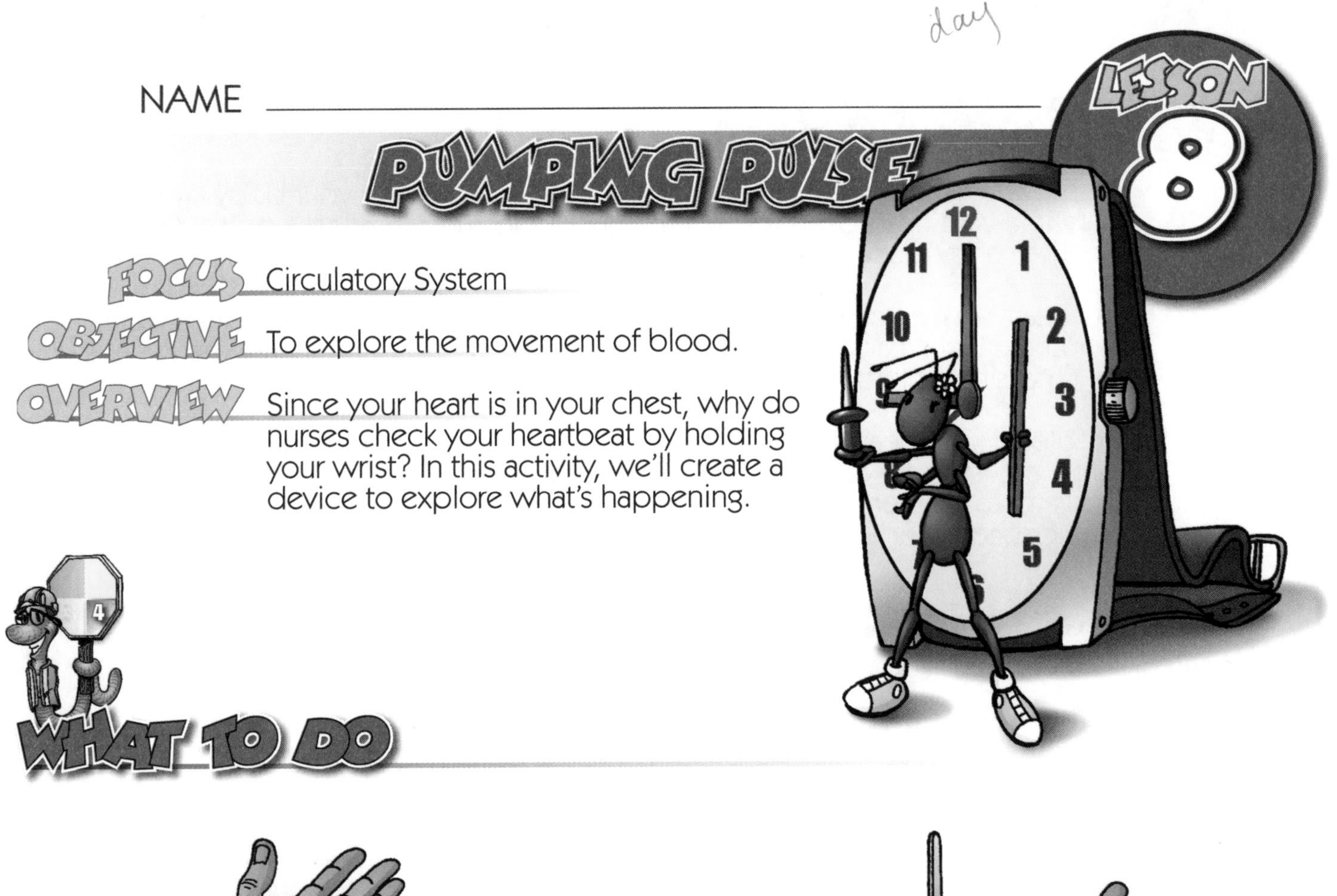

PUMPING PULSE

FOCUS Circulatory System

OBJECTIVE To explore the movement of blood.

OVERVIEW Since your heart is in your chest, why do nurses check your heartbeat by holding your wrist? In this activity, we'll create a device to explore what's happening.

WHAT TO DO

STEP 1

Place your left arm on the table with your palm facing up. Using your right hand, **press** two fingertips into the middle of your left wrist. **Move** your fingers around until you feel a bumping movement under your skin. This is your pulse. (If you can't find your pulse, ask your teacher for help.)

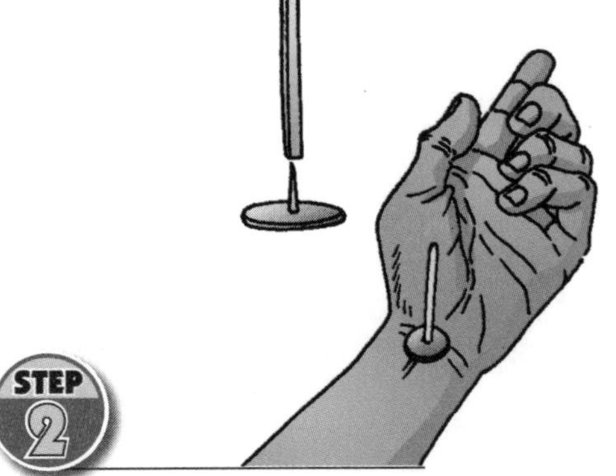

STEP 2

Carefully **push** the sharp end of the thumbtack into the bottom of the wooden match. Using the round metal thumbtack as a base, **stand** the match straight up on your arm exactly where you felt your pulse. **Watch** the match move slightly up and down. This is your pulse in action!

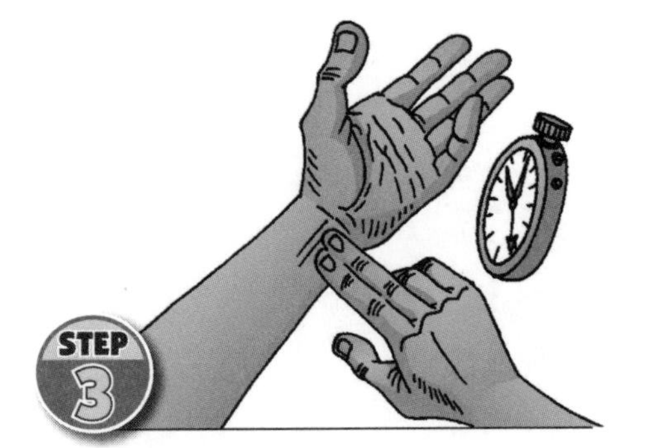

STEP 3

Ask a team member to time you for 30 seconds while you count each pulse. When they say "stop," **record** the number of pulses you've counted. **Multiply** that number by two. This is your pulse rate for one minute.

STEP 4

Hop up and down in place 50 times. Now **repeat** Step 3. **Record** the results and **compare** them with the first session. Why were the results different? **Share** and **compare** observations with your research team.

The movement you felt was **blood** being **pushed** through your body by your **heart**. Each time your heart beat, you felt a slight bump. This is called a **pulse**. (Your wrist is simply a handy place for the nurse to check your pulse.)

Your heart is a marvelous **muscle** that pumps 24 hours a day, seven days a week — for your entire life! It can pump around six quarts of blood per minute when you're resting. (That's enough to fill a railroad tank car in less than a day!)

The blood your heart pumps brings oxygen to your entire body. Your heart is constantly adjusting to your body's needs. For instance, when you exercise, your body needs more **oxygen** and **energy**. As you saw in Step 4, your heart compensates for this increased need by pumping faster.

What is a pulse? Describe what your pulse felt like in Step 1.

Compare Step 1 with Step 2. How were they similar? How were they different?

What does your heart do? Describe how it adjusts to your needs. What would happen if it stopped beating?

Compare Step 3 with Step 4. How were they similar? How were they different? What caused this difference?

Based on what you've learned, why is the heart an important muscle?

Your heart is a muscle that pushes blood through your body. The rate your heart beats is called the pulse. Pulse rates can change depending on your body's needs.

Philippians 2:5-8 Using a matchstick to watch your heart beat is a fun way to see evidence of life. Every heartbeat is pushing life-giving blood through your body.

What are you doing with the life God gave you? Does your life revolve around your own wants and needs, or do you think about others' needs and how to help them? This Scripture reminds us that Jesus' entire life was dedicated to serving others. Let Jesus be your example. Like Jesus, make your life a blessing to those around you.

My Science Notes

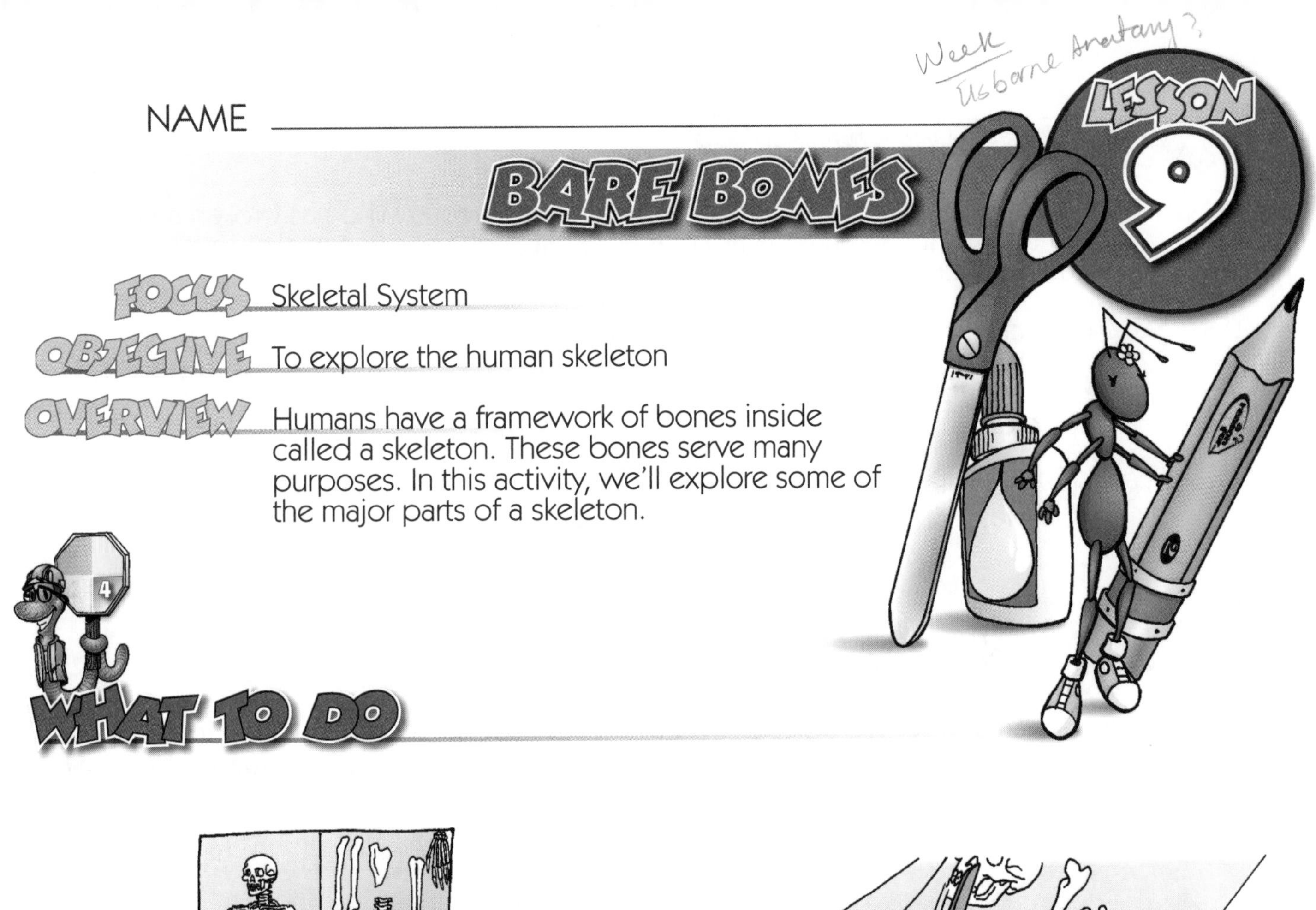

NAME ______________________________

BARE BONES

FOCUS Skeletal System

OBJECTIVE To explore the human skeleton

OVERVIEW Humans have a framework of bones inside called a skeleton. These bones serve many purposes. In this activity, we'll explore some of the major parts of a skeleton.

WHAT TO DO

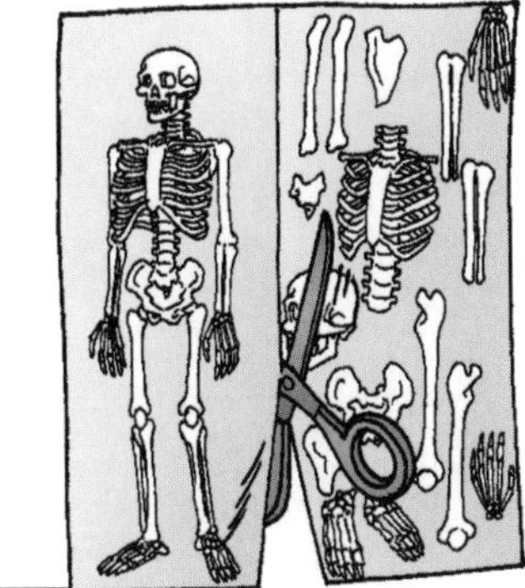

STEP 1

Remove the **Bare Bones** sheet from the back of your worktext (page 163). On one side of the sheet is a picture of a skeleton. On the other side is a bunch of bones. Carefully **cut** the sheet along the dotted line. **Set** the left half aside for now.

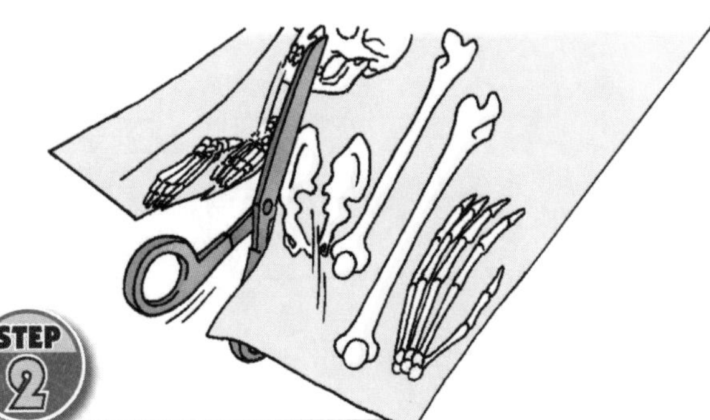

STEP 2

Pick up the right half of the worksheet and carefully **cut out** each group of bones. (The neater the cuts, the easier the next step.) **Compare** the bones, observing how they are alike and how they are different. **Sort** the bones into groups according to what you've observed.

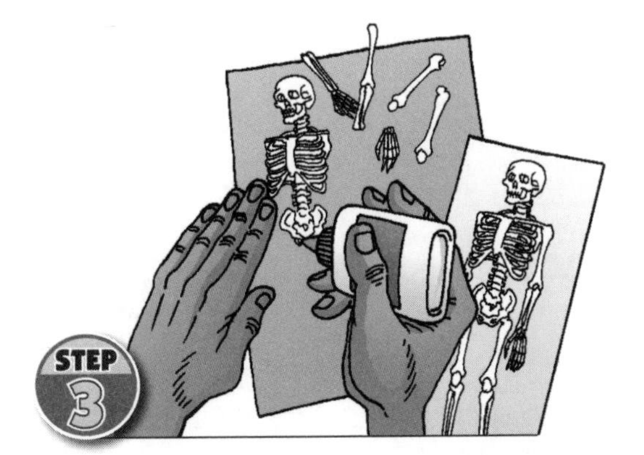

STEP 3

Take out a new sheet of paper. On the left half of the page, carefully **arrange** the bones into a human skeleton. (Use the skeleton sheet as a guide.) Once you've assembled your skeleton, **glue** the bones to the paper.

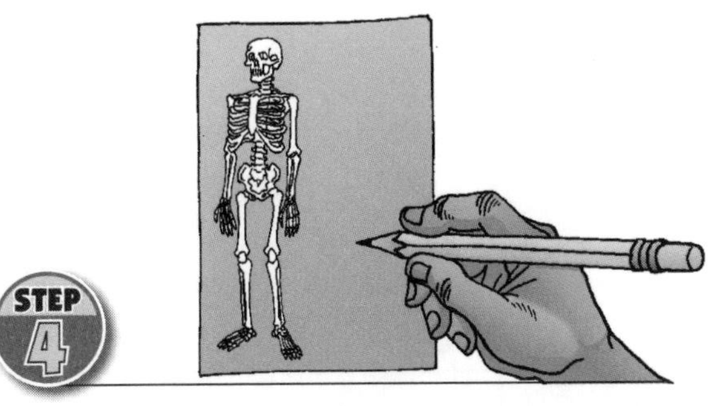

STEP 4

Draw a short line from the skull toward the right side of the paper. **Write** "skull" at the end of the line. **Draw** lines and **label** bones until your sheet is completely labeled. **Review** each step in this activity. **Discuss** what you've observed with your research team.

Did you know that your **bones** are alive? Ask someone who has broken a bone, and they'll tell you it can hurt! Compare that to clipping your fingernails. That doesn't hurt because fingernails are not alive.

Your **skeleton** is actually 206 individual bones all working together. It's much more than just a frame to hold you up! It works with other body parts (like **tendons**, **ligaments**, and **joints**) to allow you to move. It protects your vital **organs** (**heart**, **lungs**, **brain**, etc.). It protects key parts of your **nervous system** (like the **spinal cord**).

Your bones also contain **nutrients** your body needs, including calcium and phosphorous. Bones even produce special **blood cells** that help fight **disease**.

Describe the bones you cut out in Step 2. Were they individual bones or groups of bones? How many bones are there in a human skeleton?

Describe the skull. What does the skull protect? Why is this important?

Why does it hurt when you break a bone?
Why doesn't it hurt to clip your fingernails?

Describe two things bones do (in addition to holding you up).

A leg bone is large and thick compared to a finger bone. Explain why this might be. How might a bone's size and shape affect its use?

The human skeleton contains 206 individual bones. It works with other body parts to allow movement. It protects our organs and spinal cord. It even provides nutrients and helps fight disease.

Galatians 5:22 You may have started this lesson thinking a skeleton was just a bunch of dead bones. Now you understand the many ways your skeleton works — helping you move, protecting your organs, even fighting disease. You've gained a whole new way of looking at bones!

Just as your skeleton has many functions, God's spirit works in many ways, too. This Scripture talks about exciting changes that happen when the Holy Spirit controls our lives. When you understand God's love, you gain a whole new way of looking at life!

My Science Notes

Forces
CHAPTERS 10-18

Forces

In this section, you'll discover how "**push** and **pull**" form the basis for all physical movement. You'll explore simple machines (levers, pulleys). You'll work with Newton's laws of motion. You'll even learn to understand concepts like torque, inertia, and buoyancy.

NAME ______________________________________

LESSON 10

THREE STATES

FOCUS States of Matter

OBJECTIVE To explore the basic states of matter.

OVERVIEW Everything around us is made of something, even things we can't see! For example, when something dissolves it appears to just go away — but does it? In this activity, we'll find out.

WHAT TO DO

STEP 1

Using the magnifying lens, **examine** pieces of table salt. **Look** closely at their shape.

STEP 2

Fill your container half full of water. Now **stir** in small amounts of salt and **watch** it dissolve. **Continue** stirring and slowly **add** salt until no more salt will dissolve.

STEP 3

Cut black construction paper into a large circle and **place** it in the bottom of a pie pan. **Pour** the liquid from your container slowly onto the black circle until the paper is completely soaked. (Use the liquid only. Don't let any extra salt drip onto the paper.)

STEP 4

Place the pan in a warm, sunny spot and let it dry for a few days. Now using the magnifying lens, **examine** the paper closely. **Look** for things similar to what you saw in Step 1. **Share** and **compare** observations with your research team.

What are some ways that **matter** was changed in this activity? First, you **dissolved** the **solid** table salt into a **liquid** (water). The salt seemed to disappear, but it really didn't. It was broken down by the water into very tiny parts to form a **solution**.

When you poured the solution onto the paper and let it dry for a few days, the water **evaporated**, changing into a **gas**. The material remaining on the paper is the original salt — now a solid again!

Changes in matter take place around you every day! When you make a powdered drink mix, dry the laundry, or wash your hands, you're creating **physical changes** in the **state of matter**. It's the same material, just a different form.

WHAT WE LEARNED

Describe the salt crystals you looked at in Step 1.
On a sheetof paper, draw some of the shapes you saw.

Circles, squares, white

In step 2, why did you have to quit adding salt?
Why couldn't you add more?

It would not dissolve

3 In step 3, what did the liquid look like when you poured it? Could you see any salt on the paper?

No, We could not See the salt /Cloudy but
clear water

4 After the paper dried, could you see any salt on the paper? If so, where did the salt come from?

Yes, the water dried + salt was left
looks like ice and crystals, Snow

5 Give at least two examples of other common physical changes in matter (like a puddle drying up).

Matter changes form. Even though it's the same material, it may appear as a solid, a liquid, or a gas. These kinds of changes happen around us all the time.

I John 5:7 In this activity, the pan and the air surrounding it contained all three states of matter — solid, liquid, and gas. All three were matter, but they appeared in different forms!

John tells us that the Father, the Word (Jesus), and the Holy Spirit all work together to provide us with salvation. Even though all three are different, the three are one.

NAME day

LESSON 11

NEEDLE BOAT

FOCUS Surface Tension

OBJECTIVE To discover how water molecules attract each other.

OVERVIEW Everything around us is made of matter, but some kinds of matter stick together in strange ways. For example, you can make a piece of steel float if you know how! In this activity, we'll explore why.

WHAT TO DO

STEP 1

Fill your container with water. Make certain it is almost completely full.

STEP 2

Using the tweezers, **pick up** the needle. Be sure to **grab** it in the center.

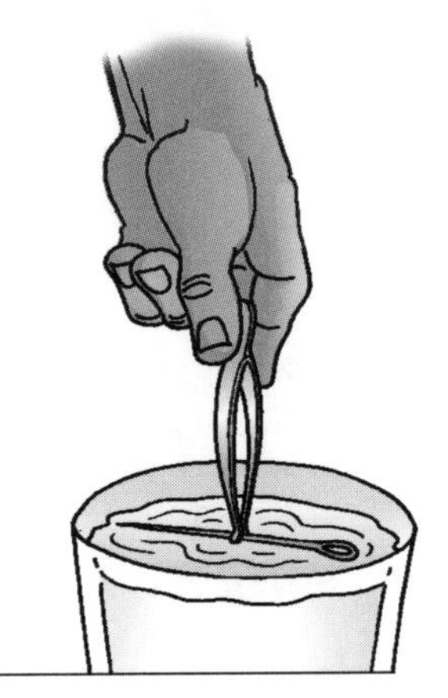

STEP 3

Slowly and carefully **place** the needle on the surface of the water. (This may take a bit of practice, so don't get discouraged.) If the needle sinks, just try again!

STEP 4

Look closely at the floating needle and the surface of the water. Carefully **examine** the area around the needle. **Discuss** what you see with your research team.

What Happened?

Water molecules stick together because of their **structure**. We call this tendency **surface tension**. Imagine that the water molecules at the surface are holding hands.

When the needle was placed gently on the water, the little hands were strong enough to hold it up. Their combined **force** was even stronger than **gravity** trying to **pull** the needle down. (The **interaction** between these two forces makes the dent you see in the water around the needle.)

Surface tension is what makes raindrops round. It makes water bead on a window, and it makes the surface bend along the edge of a bucket of water. But remember, surface tension is not a strong force. A needle can't float on its point because there are only a few water molecules to support it!

What We Learned

1. **Is the surface of a container of water completely flat? What happens along the edges? Describe the shape of the surface.**

2. **Why do you think it was important in Step 2 to grab the needle in the center? Why wouldn't it work to hold it by one end?**

In Step 3, did the needle float the first time? If not, what did you do differently to succeed?

What do you think the water molecules do that helps the needle float?

Using what you've learned in this lesson, describe why it's easier to dive head-first into a pool than to fall flat on the surface? (Ouch!)

CONCLUSION

Surface tension shows how some kinds of matter (in this case, water molecules) tend to stick together. Understanding this can help us not only float needles, but also float huge ships!

Romans 12:4-5 In this Scripture, Paul talks about the importance of working together for the good of all. Just as water molecules must "hold hands" to make the needle float, so we must stand together as God's children.

Even though we are individuals with different skills and abilities, together we can make a difference in our school and in our world!

My Science Notes

NAME ____________________ day

LESSON 12

GREAT DECEIVER

FOCUS Absorption

OBJECTIVE To explore how water is absorbed by materials.

OVERVIEW Water is one of the most common materials on Earth. Many things we do each day involve using, moving, or storing some form of water. In this activity, we'll explore one method of water storage.

DEHYDRATED GEL

WHAT TO DO

Number three paper cups — 1, 2, 3. **Pour** one capful of dehydrated gel into Cup 1.

Pour the gel from Cup 1 into Cup 2. Now **pour** the gel back into Cup 1. **Observe** how the material moves from one cup to the other. Before beginning Step 3, **set** Cup 1 (with the gel) aside.

Pour about one inch of water into Cup 3. Now **pour** the water into Cup 2, then back into Cup 3. **Observe** how the water moves from one cup to the other.

Pour the water from Cup 3 into the gel in Cup 1. **Count** to 50 slowly. Now **pour** the material from Cup 1 into Cup 2. What happened? **Look** inside Cup 1 and **discuss** what you find with your research team.

What Happened?

The **dehydrated gel** used in this activity is very **hygroscopic** — a big word that means it can soak up a lot of water! This gel actually can soak up and hold many times its own **weight** in **water**.

The **absorbing** ability of dehydrated gel makes it perfect for many tasks. One of the most common uses is to make disposable diapers!

Dehydrated gel is also helpful in transporting live plants over long distances. The gel prevents water from **evaporating**, making water available to the plant roots for a much longer period of time.

What We Learned

Describe the gel in Step 1. What is it similar to? What state of matter is this?

____

____

____

____

____

____

____

Describe how the gel poured in Step 2. What makes it pour so easily?

____

____

____

____

____

____

____

3 Describe how the water poured in Step 3. How did it pour differently from the gel? What state of matter is water?

4 Describe how the gel reacted after Step 4. What do you think happened?

5 Give at least three examples of water absorption.

Absorption is the attraction and holding of liquid in a material. We use absorption in many ways — like drying our hands, mopping the floor, or even handling huge oil spills.

Luke 8:12 Things aren't always what they appear. You were probably fooled by the demonstration the teacher did to introduce this lesson! But once you understood what was happening, you knew the truth.

Some people call the devil the "great deceiver." What they mean is that he's always trying to fool us into believing things that aren't true! The devil doesn't want us to believe that God loves us, or that Jesus died to save us. But if we stay close to Jesus, we won't be deceived by Satan's lies.

My Science Notes

NAME ______________________

GRAVITY STOPPER

LESSON 13

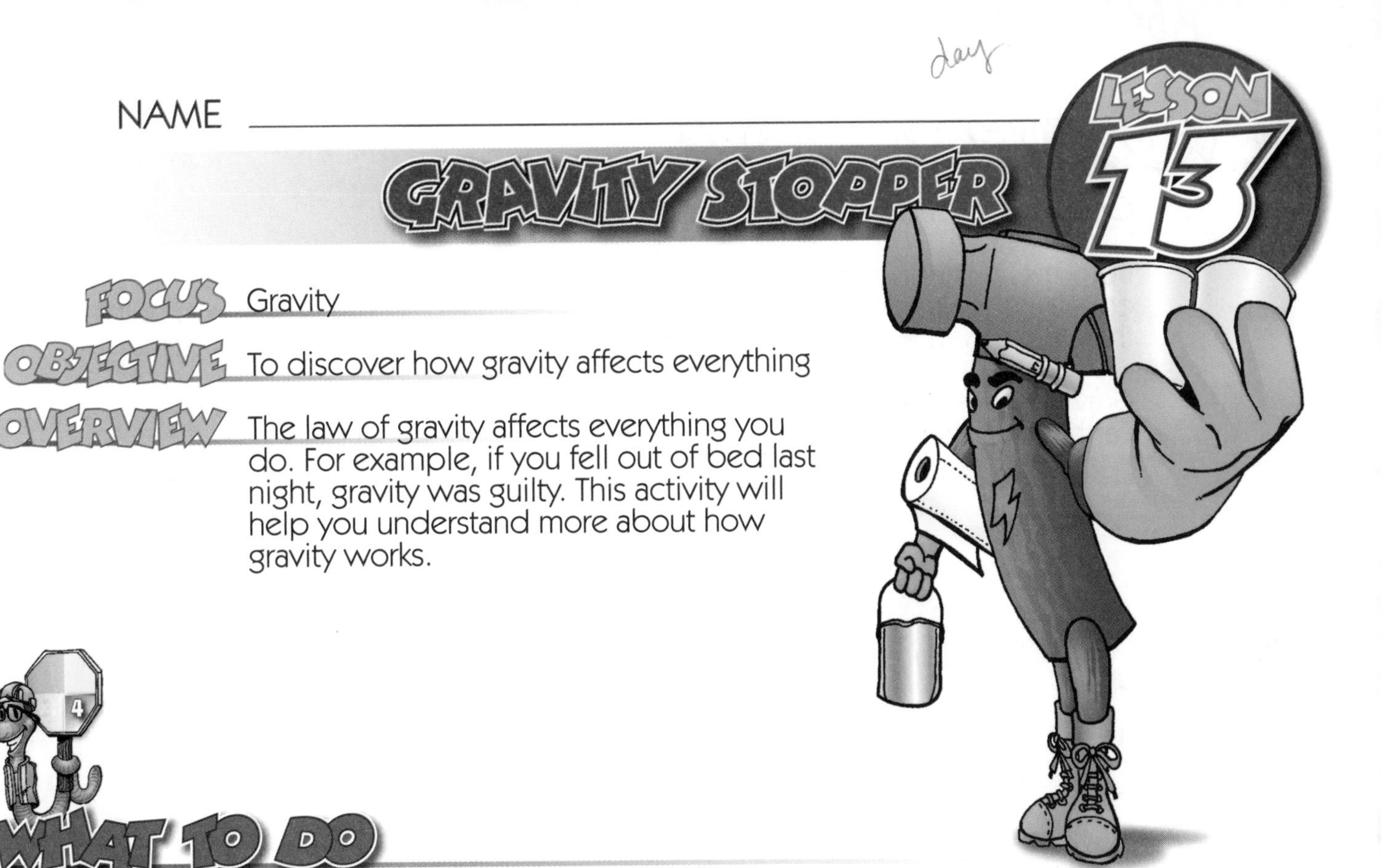

FOCUS Gravity

OBJECTIVE To discover how gravity affects everything

OVERVIEW The law of gravity affects everything you do. For example, if you fell out of bed last night, gravity was guilty. This activity will help you understand more about how gravity works.

WHAT TO DO

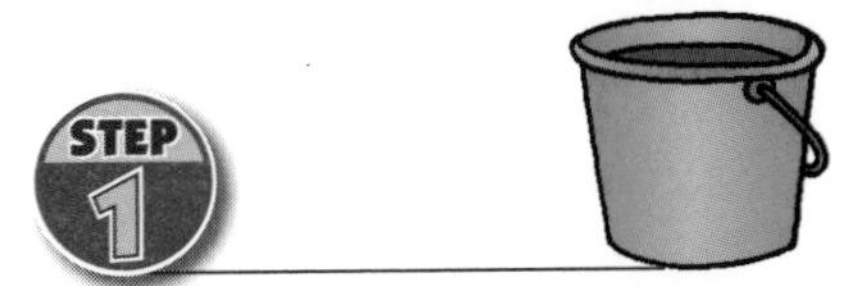

STEP 1

Fill a paper cup with water. **Hold** the cup about three feet above an empty bucket.

STEP 2

Ask the members of your research team to stand at one side and **observe** closely as you **drop** the cup straight down into the bucket.

STEP 3

Make a hole in the bottom of the cup with a sharp pencil. **Hold** your finger over the hole and **refill** the cup with water. **Ask** team members to **observe** closely as you **hold** the cup over the bucket again. **Remove** your finger from the hole, letting the water run out until the cup is empty.

STEP 4

Hold your finger over the hole and **refill** the cup with water. **Ask** team members to **observe** closely as you **hold** the cup over the bucket. **Remove** your finger from the hole and immediately **drop** the cup into the bucket. **Discuss** the water's behavior in each step.

Once again, the **law of gravity** has been upheld!

Both the cup and the water obeyed the law of gravity. Gravity **pulls** on everything equally. In Step 4, the water quit coming out of the hole because it couldn't fall faster than the cup. They were both being pulled by the same force — gravity!

Gravity is what makes raindrops fall, makes rivers run, and keeps us from floating off the Earth into space. Water, cups, and even people all have to obey the law of gravity.

In Step 1, why didn't the cup fall into the bucket? What force was trying to make the cup drop?

In Step 2, what happened when you let go of the cup?

3 In Step 3, what force pulled the water out of the cup? What kept the cup from falling?

4 In Step 4, what happened to the water when you let go of the cup and removed your finger from the hole at the same time?

5 Using what you've learned in this lesson, describe what might happen to us if there were no more gravity.

Whether moving, standing still, or falling — everything in our world obeys the law of gravity. No machine or device ever made can move without taking gravity into consideration.

Joshua 6:20 One of the best examples of gravity in Scripture is found in the story of Joshua and the battle of Jericho. When the trumpets blew and the people shouted, the walls of the great city came tumbling down!

Just as every object in the universe obeys the law of gravity, all believers should obey the laws of God. Since God made us and knows what is best for us, doesn't it make sense that God's laws are best?

My Science Notes

NAME ________________

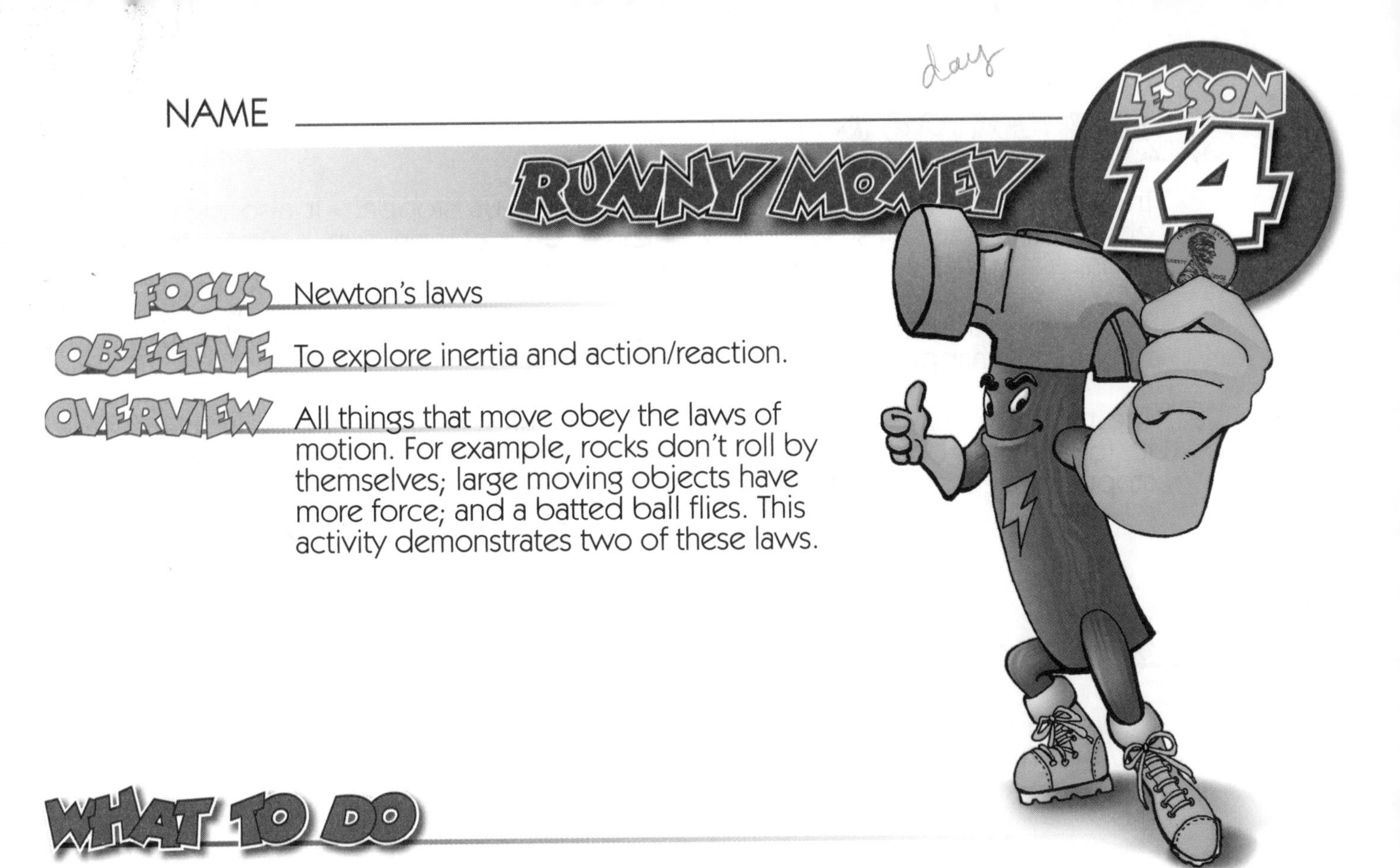

FOCUS Newton's laws

OBJECTIVE To explore inertia and action/reaction.

OVERVIEW All things that move obey the laws of motion. For example, rocks don't roll by themselves; large moving objects have more force; and a batted ball flies. This activity demonstrates two of these laws.

WHAT TO DO

For this activity, you will need six pennies. **Stack** five of the pennies on top of each other. Make sure the stack is straight.

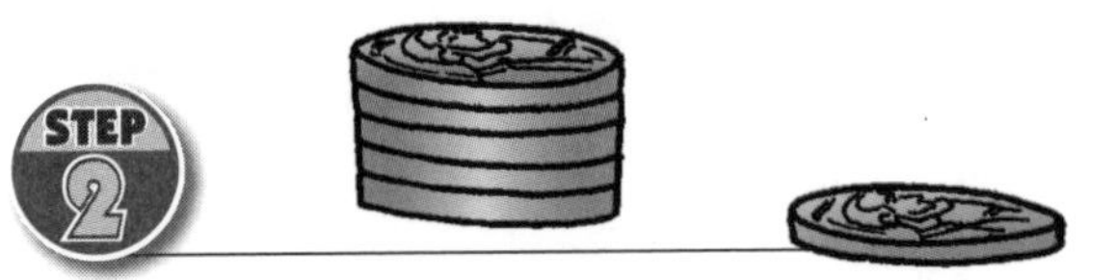

Place the sixth penny a few inches away from the stack. **Predict** what might happen if you flick the penny into the bottom of the stack.

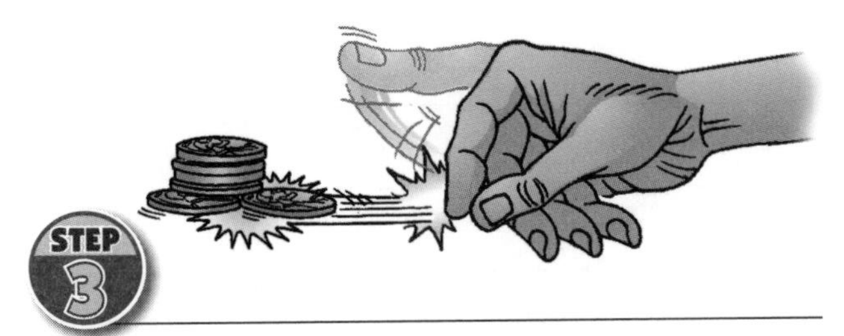

Flick the penny into the stack. **Make** the penny hit hard and fast, and try to hit the stack head on. (This may take a bit of practice, so don't get discouraged.) If you miss, **move** the penny a bit closer and try again!

Repeat Step 3 until everyone has had a turn. **Observe** what happens when the penny hits the stack each time. **Discuss** the results with your research team.

Inertia means that an object that is stopped stays stopped. It also means that an object that is moving stays moving, unless a **force** acts on it. In this case, the stack of pennies stayed put until they were struck by the moving penny.

Action/Reaction means that if a force creates an action, this action will cause something else to happen.

The moving penny (action) stopped when it hit that stack because it **transferred** its force to the bottom penny. The bottom penny moved (reaction), but eventually stopped because of **friction**.

In Step 1, once you made the stack, what kept it from moving around? What law is this?

In Step 2, what did you predict would happen when the penny hit the stack? Why?

3 Why do you think it was important to hit the stack "head on" in Step 3?

What was the "action" in Step 4? What was the "reaction"? How was inertia involved?

Using what you've learned in this lesson, describe why it's important to always wear a seatbelt.

Conclusion

All objects obey the laws of motion. Inertia and Action/Reaction are two of Newton's three laws of motion.

Food for Thought

Matthew 11:28-29 Sometimes we have a lot in common with that bottom penny. Our problems seem to keep stacking up on top of us until we can't move. They get heavier and heavier, holding us down so we feel we can never escape.

Then Jesus comes along and applies a powerful force to our lives! Just as the moving penny replaces the trapped penny, Jesus takes our place and sets us free. Since Jesus has taken such a powerful action, what should your reaction be?

Journal My Science Notes

NAME ______________________

TEETER TOTTER PENNIES

LESSON 15

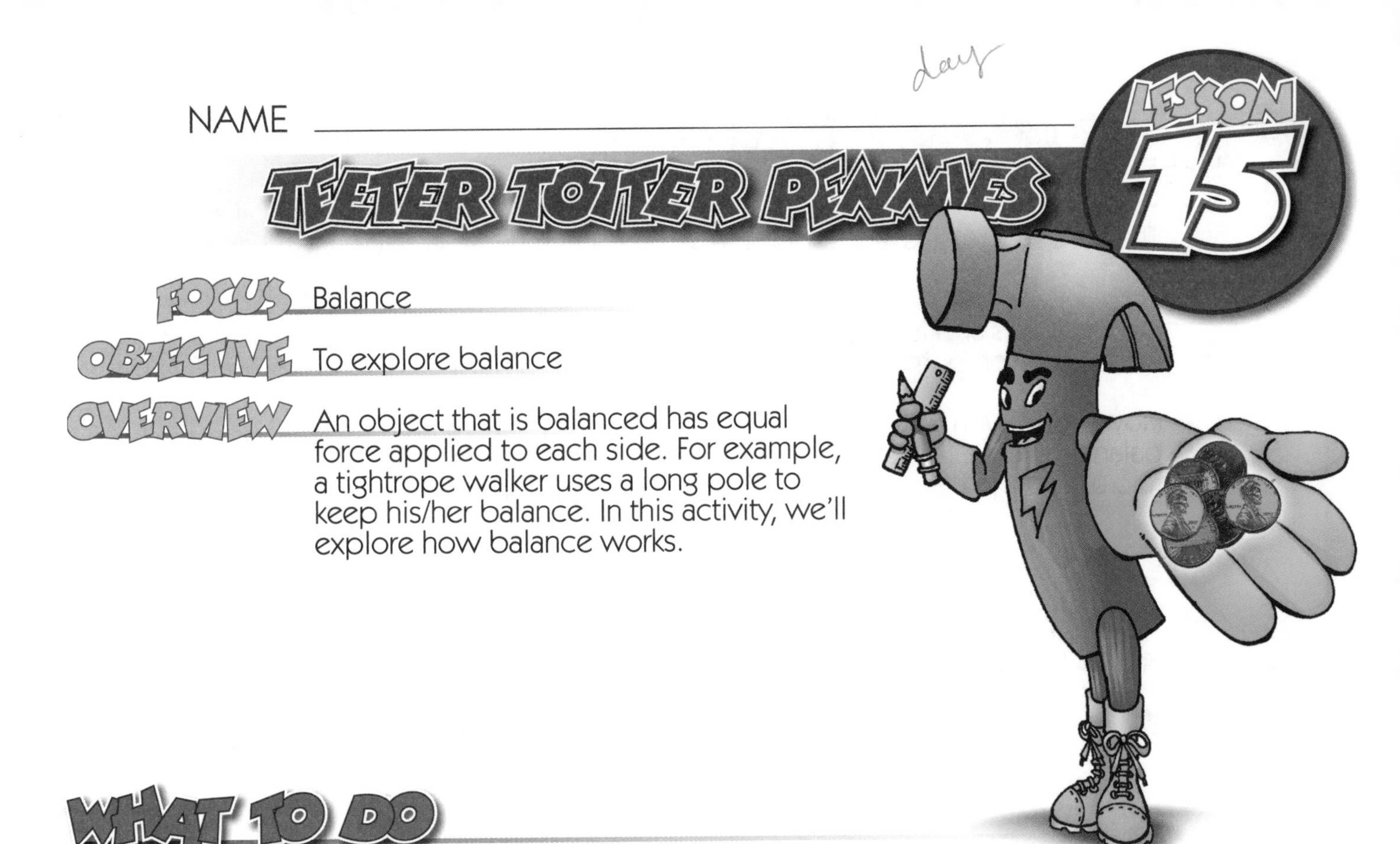

FOCUS Balance

OBJECTIVE To explore balance

OVERVIEW An object that is balanced has equal force applied to each side. For example, a tightrope walker uses a long pole to keep his/her balance. In this activity, we'll explore how balance works.

WHAT TO DO

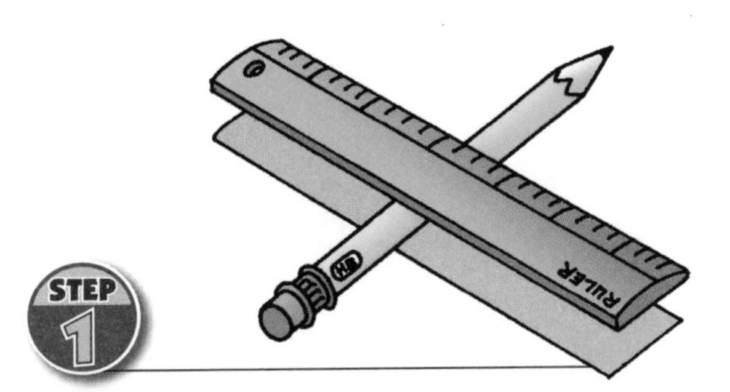

Place a round pencil on a flat table or desk. **Balance** a ruler across the pencil so that the ruler is level and motionless. **Note** the spot where the pencil touches the ruler. This is the "balance point."

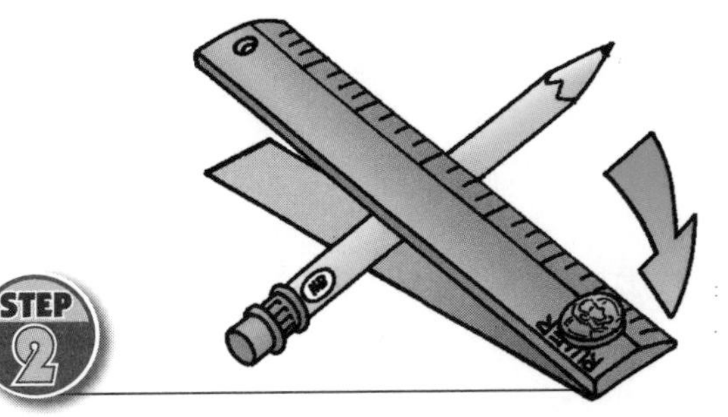

Place a penny on one side of the ruler. **Discuss** what happens to the ruler. (Make sure the pencil is still touching the balance point.)

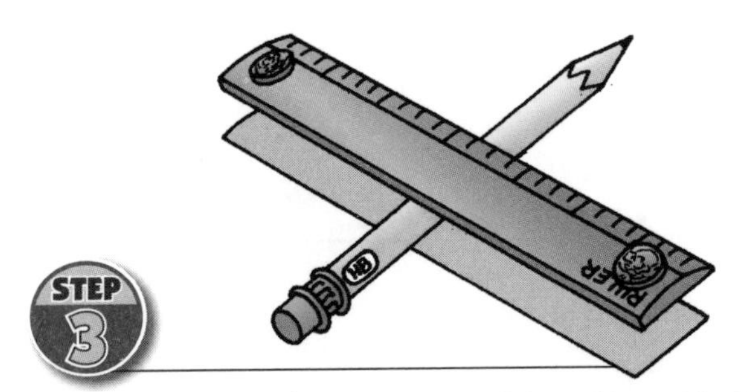

Place a penny on the other side of the ruler and **balance** the ruler again. (Make sure the pencil is still touching the balance point.) **Discuss** what you see happening.

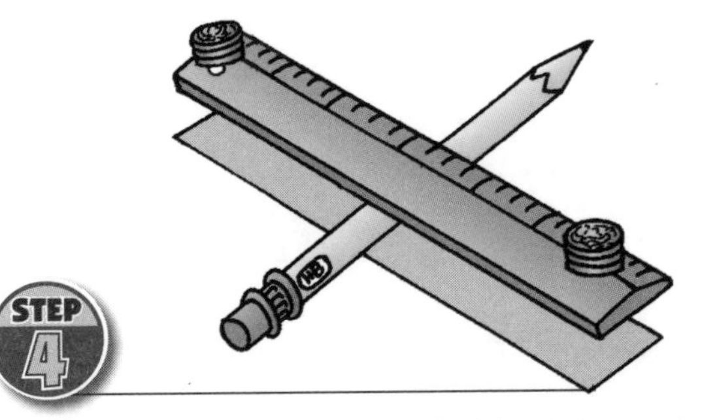

Add two more pennies to each side. **Balance** the ruler again. (Make sure the pencil is still touching the balance point.) Now with your research team, **discuss** what happened in each step.

WHAT HAPPENED?

You had to apply equal **force** (weight) to each side of the ruler to make it **balance**. The balance point is also called a **fulcrum**. When the ruler is balanced on the fulcrum, it acts as a **lever**.

We use levers and fulcrums in different ways for different tasks. Tightrope walkers need equal force on both sides of their long poles (a form of lever), or they will fall!

However, if you're opening a paint can with a screwdriver (lever) you don't want balanced force, or the lid won't open! More force is required on one side of the fulcrum (in this case, the rim of the can) for the process to work.

In Step 1, how much of the ruler was on each side of the pencil when the ruler was balanced?

In Step 2, why did the ruler tilt to one side? Predict what you might do to balance it again.

Compare the location of the second penny added in Step 3 to the first penny added in Step 2. How do they relate to the balance point?

4. How did adding more pennies in Step 4 affect the balance of the ruler? What did you have to keep in mind?

5. Using what you've learned in this lesson, describe where four equal-size kids would need to sit to make a teeter-totter operate best.

Balance is a result of equal force applied to both sides of a lever. Using levers and fulcrums, we use balance in many ways to make our work easier.

Matthew 14:23 Scripture tells us that Jesus always made special time to be alone with God. This is what helped him keep his life in balance no matter what happened.

But sometimes we get so busy with school and sports and music and hobbies and all sorts of other things, that we forget about God. Leaving God out of our lives will eventually make us unbalanced. This leads to all sorts of problems! Fortunately, God never forgets about us. Ask your teacher to help you discover ways to spend more time with God. Begin now to find balance for the rest of your life!

My Science Notes

NAME ____________________

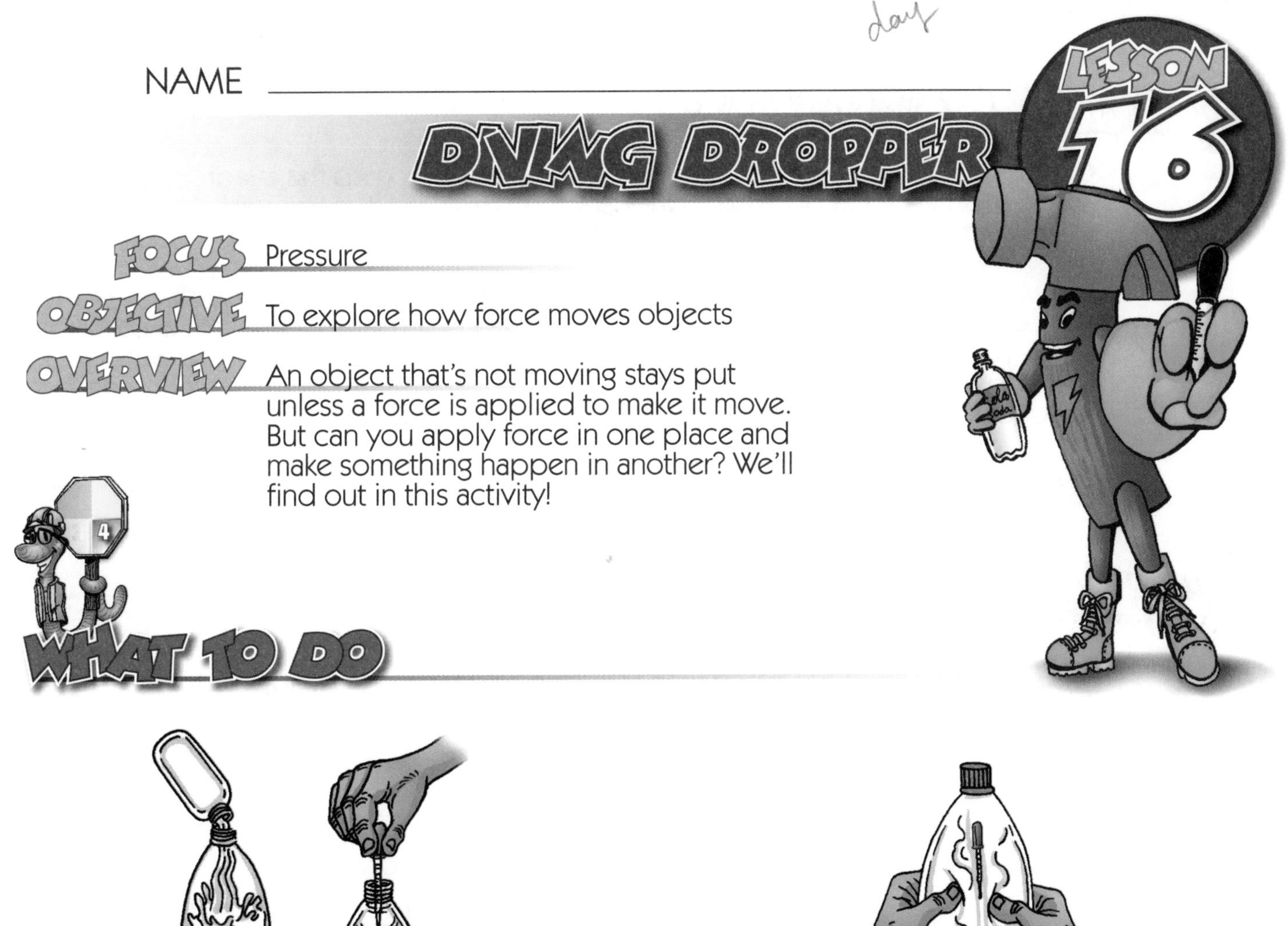

Diving Dropper

Focus Pressure

Objective To explore how force moves objects

Overview An object that's not moving stays put unless a force is applied to make it move. But can you apply force in one place and make something happen in another? We'll find out in this activity!

What to Do

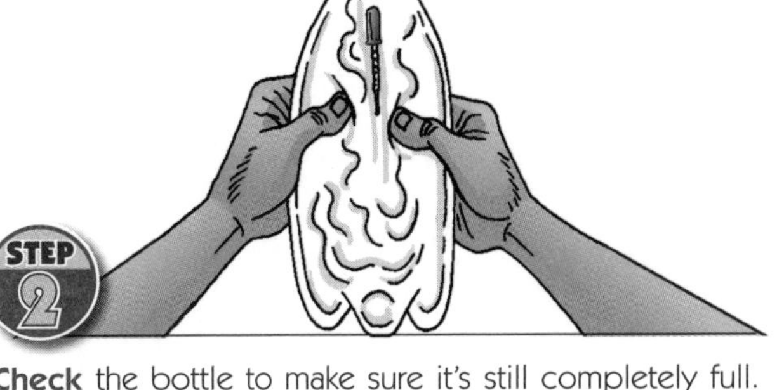

Step 1

Completely **fill** a two-liter bottle with water. **Place** the tip of a glass eyedropper in the water, then **squeeze** and **release** the bulb. The eyedropper should pick up enough water to be almost, but not quite, full. (Make sure there's a small air bubble inside.) Gently **place** the eyedropper in the bottle. It should float just below the surface.

Step 2

Check the bottle to make sure it's still completely full. If not, **add** water until it is. Tightly **fasten** the lid back on the bottle. (Make sure it's really tight!) Now, **hold** the bottle upright and firmly **squeeze** it.

Step 3

Watch to see what happens to the eyedropper. If nothing happens, **squeeze** the bottle a little harder. (Note: If there's still no change, **open** the bottle and **place** a little more water in the eyedropper. **Replace** the eyedropper. Make sure the bottle is completely full, then **try again**.)

Step 4

Squeeze the bottle again and **observe** what happens. Now let go of the bottle and **observe** what happens. **Watch** the air bubble in the dropper closely as you squeeze and release the bottle. **Review** each step and **discuss** your observations with your research team.

Gravity is always trying to **pull** things down. In order to keep the eyedropper floating, another **force** had to **oppose** (fight) gravity. In this case, the trapped air bubble provided **buoyancy** (floating) to oppose gravity and keep the eyedropper floating.

But when you squeezed the bottle, you added another force! Since **liquids** (like water) can't **compress**, but **gases** (like air) can, change could only occur in the eyedropper. When you squeezed the bottle it compressed the bubble, letting more water in the eyedropper. This made the eyedropper heavier, so it sank! Releasing the force made the bubble **expand**, pushing out water so that the eyedropper could float again.

Applying force in one place caused something to happen in another. Scientists call this the "**transfer of forces**."

WHAT WE LEARNED

In Step 1, what force was trying to pull your eyedropper down? Why didn't the eyedropper sink?

Why was it important to make sure the lid was screwed on tightly in Step 2?

What happened when you squeezed the bottle in Step 3?
Where did the force that made the dropper sink come from?

In Step 4, what happened when you stopped squeezing the bottle (stopped applying force)? Why?

Using what you learned in this activity, explain how submarines dive and surface. Would a sub need more or less water in its tanks to dive? What would it need to do to surface?

Conclusion

Forces move things. Planes fly, cars drive, submarines dive . . . all thanks to different forces making them move. Force can also be transferred from one place to another.

Food for Thought

Mark 9:23; 10:27 Sometimes we face things that seem impossible — like schoolwork, a big project, or a situation at home. But just as we put our force into the bottle to help the eyedropper dive, God can put his force inside us to help us accomplish what seems impossible!

How can this happen? Spend time each day getting to know God better, and soon you'll learn to trust Him with all your needs. Get connected with God, the source of all power, and dive into life!

Journal **My Science Notes**

NAME ______________________________

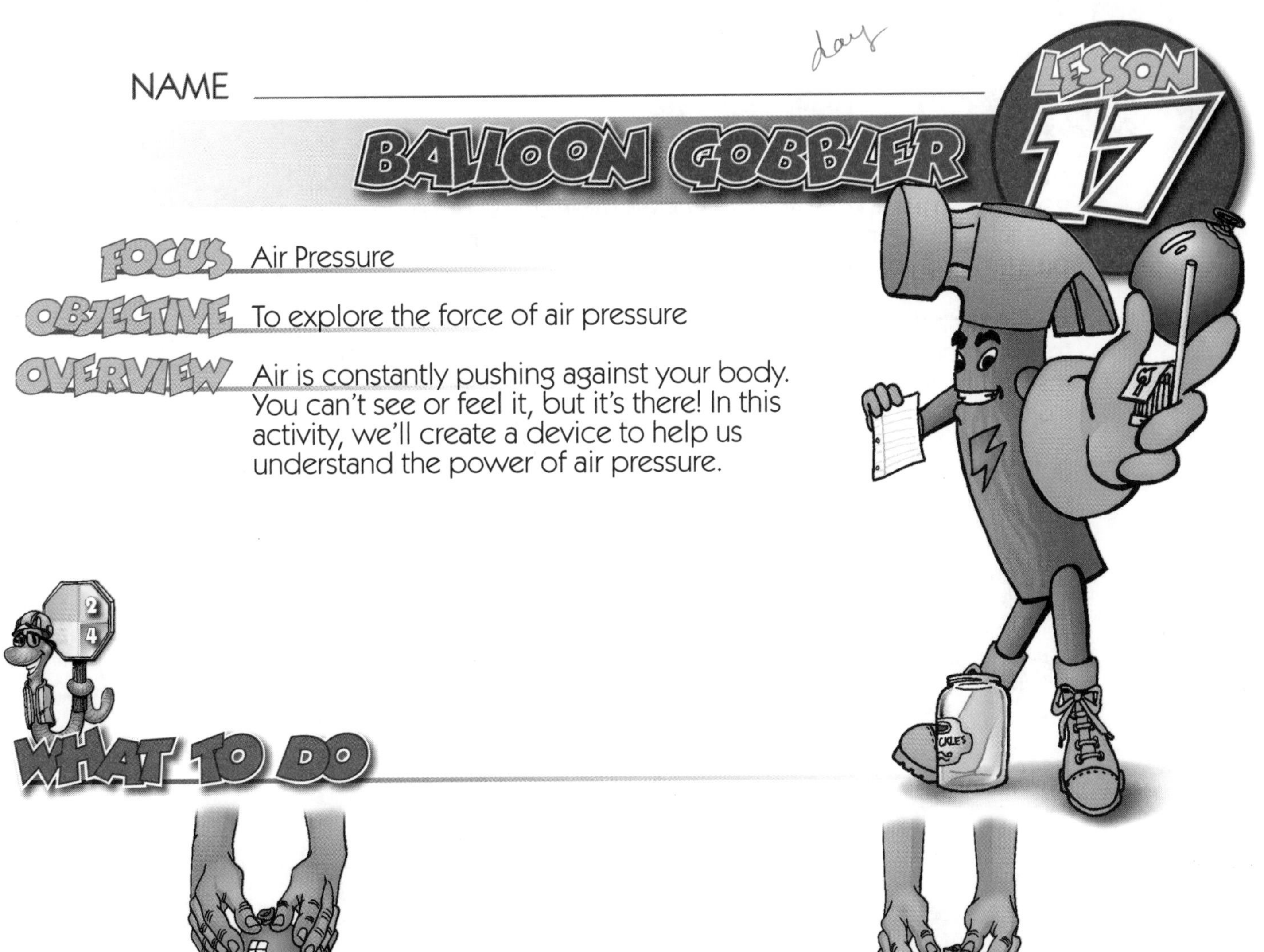

LESSON 17

BALLOON GOBBLER

FOCUS Air Pressure

OBJECTIVE To explore the force of air pressure

OVERVIEW Air is constantly pushing against your body. You can't see or feel it, but it's there! In this activity, we'll create a device to help us understand the power of air pressure.

WHAT TO DO

STEP 1

Set a wide-mouth jar on your work surface. **Fill** a balloon with water until it's slightly bigger than the jar's opening. Try to gently **push** the balloon into the jar. **Record** the results.

STEP 2

Crumple a piece of paper and **place** it in the jar. **Watch** as your teacher sets the paper on fire. As soon as the paper is burning well, quickly **place** the balloon back on the jar's mouth.

STEP 3

Watch what happens to the balloon. **Discuss** why this might have happened. **Record** your observations.

STEP 4

Try to **remove** the balloon from the jar. **Record** the results. Now **slip** a straw past the balloon into the jar and try again. **Record** the results. **Share** and **compare** your observations with other research teams.

Air is made of tiny invisible particles called **molecules**. In Step 1, the balloon wouldn't go in because the jar was full of air molecules. **Burning** the paper in Step 2 caused the air to expand, pushing a lot of molecules past the balloon and out of the jar. (You probably saw them shaking the balloon.)

The fire used all the **oxygen** in the jar and went out. As the remaining air began to cool, the molecules began to **contract** (squeeze together). This made the **air pressure** inside the jar lower than the air pressure outside. The combination of low pressure inside **pulling**, and high pressure outside **pushing**, forced the balloon into the jar.

But the lower air pressure in the jar trapped the balloon! You couldn't pull it back out until you slipped the straw past it, allowing pressure in the jar to **equalize** again.

Why wouldn't the balloon go into the jar in Step 1? What was holding it out?

____

____

____

____

____

____

____

How did the burning paper affect the air pressure in the jar? Describe the balloon's behavior as the fire burned.

____

____

____

____

____

____

____

3 Describe what happened to the balloon in Step 3. Why did this occur?

4 Explain how the straw helped you remove the balloon from the jar in Step 4.

5 When you pour liquid from a full bottle, it doesn't pour smoothly. When the bottle is nearly empty, the flow becomes smooth. Based on what you learned, explain why this happens.

Air is made of tiny particles called molecules. Air molecules take up space, and are constantly pushing or pulling on everything. Scientists call the push/pull action of air molecules "air pressure."

Matthew 19:25-26 When you first tried to push the balloon into the jar, it seemed impossible. How could such a big balloon fit through that small opening? But you trusted your teacher to show you a way, and the balloon popped through without your pushing at all!

In this Scripture, the disciples are worried. Looking at their own efforts, they can't see how it's possible for anyone to be saved. But Jesus reminds them that the answer is to trust God. The impossible doesn't happen because of what we try to do, but from relying totally on God's power!

My Science Notes

NAME ______________________________

LESSON 18

SPINNING WING

FOCUS Torque

OBJECTIVE To explore how forces change direction

OVERVIEW Many devices help change the direction of a force. For instance, if you push down on one end of a teeter-totter, the other end goes up! In this activity, we'll explore another way force changes directions.

WHAT TO DO

STEP 1

Carefully **examine** the **Spinning Wing**. **Make notes** in your journal about the various parts of this device. Pay special attention to their shapes.

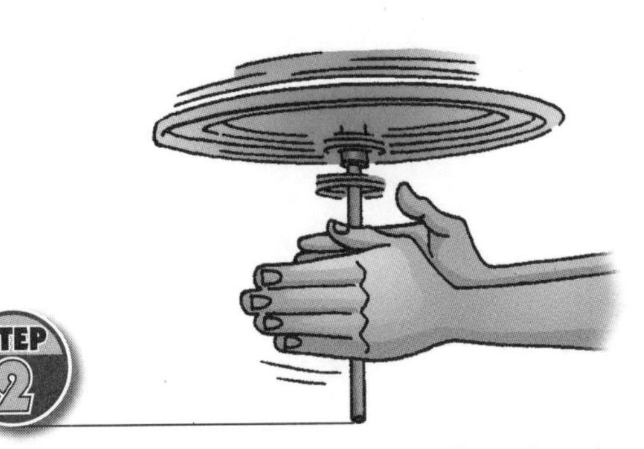

STEP 2

Hold the stick between the palms of your hands with the wing on top. Now slowly **rub** your hands back and forth. **Observe** the motion of the **Spinning Wing** and **make notes** about what you see.

STEP 3

Hold the stick again (as in Step 2). Quickly **move** your hands once in opposite directions, releasing the stick. **Record** the results. Now **repeat**, but this time reverse the direction of your hands. **Record** the results.

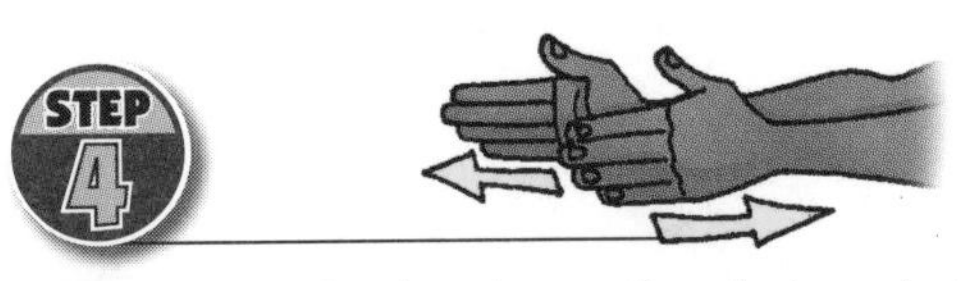

STEP 4

Repeat Step 3 using the motion that made the **Spinning Wing** fly. Give each team member three trys, then **record** the results. **Share** and **compare** observations with your research team.

What Happened?

The shape of the wing is what makes the **Spinning Wing** fly. When you made the wing spin, its shape caused the air to go over the top faster than under the bottom. Faster air has lower **air pressure**, so the wing was **pulled** upward. Scientists call this upward force **lift**.

So how was the direction of **force** changed? Your hands moved back and forth, but the stick changed the direction to round and round, then the wing changed the direction to up!

The twisting force you used is called **torque**. The more torque you provided, the longer your **Spinning Wing** flew. Torque is an important force in many modern devices.

What We Learned

1. **Describe your observations from Step 1. How were the two sides of the wing different? Where did the stick fasten to the wing? Why do you think this was important?**

2. **How were the two parts of Step 3 similar? How were they different? Which way did your right hand need to move (toward your body or away) in order to make the Spinning Wing fly?**

What force caused the Spinning Wing to spin? What force caused it to move upward? What force was trying to pull it down?

Where did the energy to make the Spinning Wing fly come from? When you moved your hands faster (providing more torque), what did the Spinning Wing do?

Based on what you've learned, describe at least two other devices that use torque (twisting force).

All movement requires some kind of force. Twisting force is called torque. Many devices are designed to cause force to change directions.

II Corinthians 5:17 In this activity, you discovered how a force can change directions. You moved your hands one way, but the **Spinning Wing** moved in another.

This Scripture talks about a change of direction that's caused by the force of God's love. The apostle Paul says that when your heart is filled with God's love, it changes the direction of your life so much that you're like a brand-new person!

My Science Notes

Earth
CHAPTERS 19-27

Earth Science

Earth Science is the study of **earth** and **sky**. In this section, you'll explore the structure of our planet (rocks, crystals, volcanos), the atmosphere (air, clouds), and related systems (water cycles, air pressure, weather). Grades 7 and 8 reach out even further with a look at the solar system and stars.

NAME ____________________

Lesson 19

Reluctant Water

Focus Matter

Objective To discover that air is another form of matter

Overview Everything around us is made of matter. Some forms are easy to see. But what about things you can't see? Are they matter too? In this activity, we'll find out!

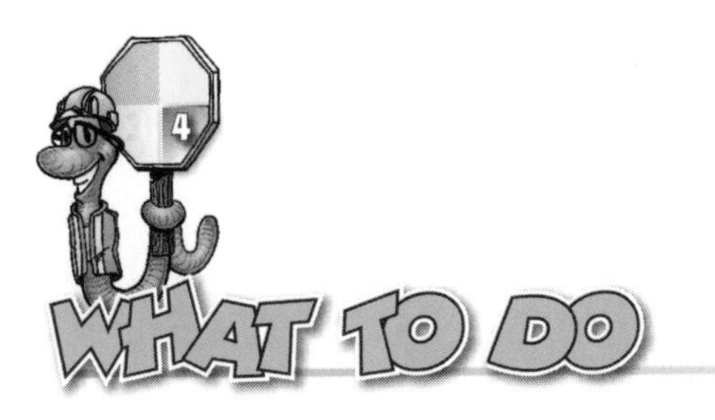

What to Do

Step 1

Place two bottles on your work surface. **Push** the two-hole stopper into one bottle. **Push** the one-hole stopper into the other bottle. **Check** to make sure the stoppers fit snugly.

Step 2

Wet the neck of the funnel. Gently **push** the funnel into the two-hole stopper. Now slowly **pour** water into the funnel. **Record** the results.

Step 3

Remove the funnel from the two-hole stopper. **Wet** the neck of the funnel, then gently **push** it into the one-hole stopper. **Check** to make sure the stopper is still tight.

Step 4

Slowly **pour** water into the funnel in the one-hole stopper. **Record** the results. Now **review** each step in this activity. **Share** and **compare** observations with your research team.

What Happened?

Are there **forms** of **matter** that you can't see? Yes! Even though you couldn't see anything in the bottle, it was full of the gaseous matter we call **air**.

Trying to add **water** to the bottles showed the presence of this invisible matter. In Step 2, the water flowed smoothly into one hole while the air came out the other. But in Step 4, there was no way for the air to get out. The water had a hard time getting in because invisible matter (air) was in the way!

Earth's **atmosphere** (air) is constantly **pushing** and **pulling** everything on the planet. This constant movement of matter even shoves around large air masses full of energy, creating what we call weather!

What We Learned

In Step 1, why was it important for the stoppers to fit snugly? What form of matter was in the bottles?

Describe what happened in Step 2. What happened to the water? How did the air leave the bottle?

Describe what happened in Step 4. What happened to the water? Why couldn't the air leave the bottle?

Based on what you've learned, explain how you know that matter can be invisible.

Give two examples of air moving something.

Everything around us is made of matter — even some things we can't see! Air is a form of matter. Air takes up space and can move. It also can move other things.

Matthew 13:15, 16 In this activity we discovered what happens when two forms of matter try to fit into the same space. The bottles couldn't be full of air and full of water at the same time. There's no room for both inside.

This Scripture reminds us that we have something in common with those bottles. When we're "full of ourselves," there's not much room for God inside. But when we open our hearts, God's love can come rushing in. When God fills our hearts with love, there's no room for selfish things any more!

My Science Notes

NAME ______________________

LESSON 20
POTATO STABBER

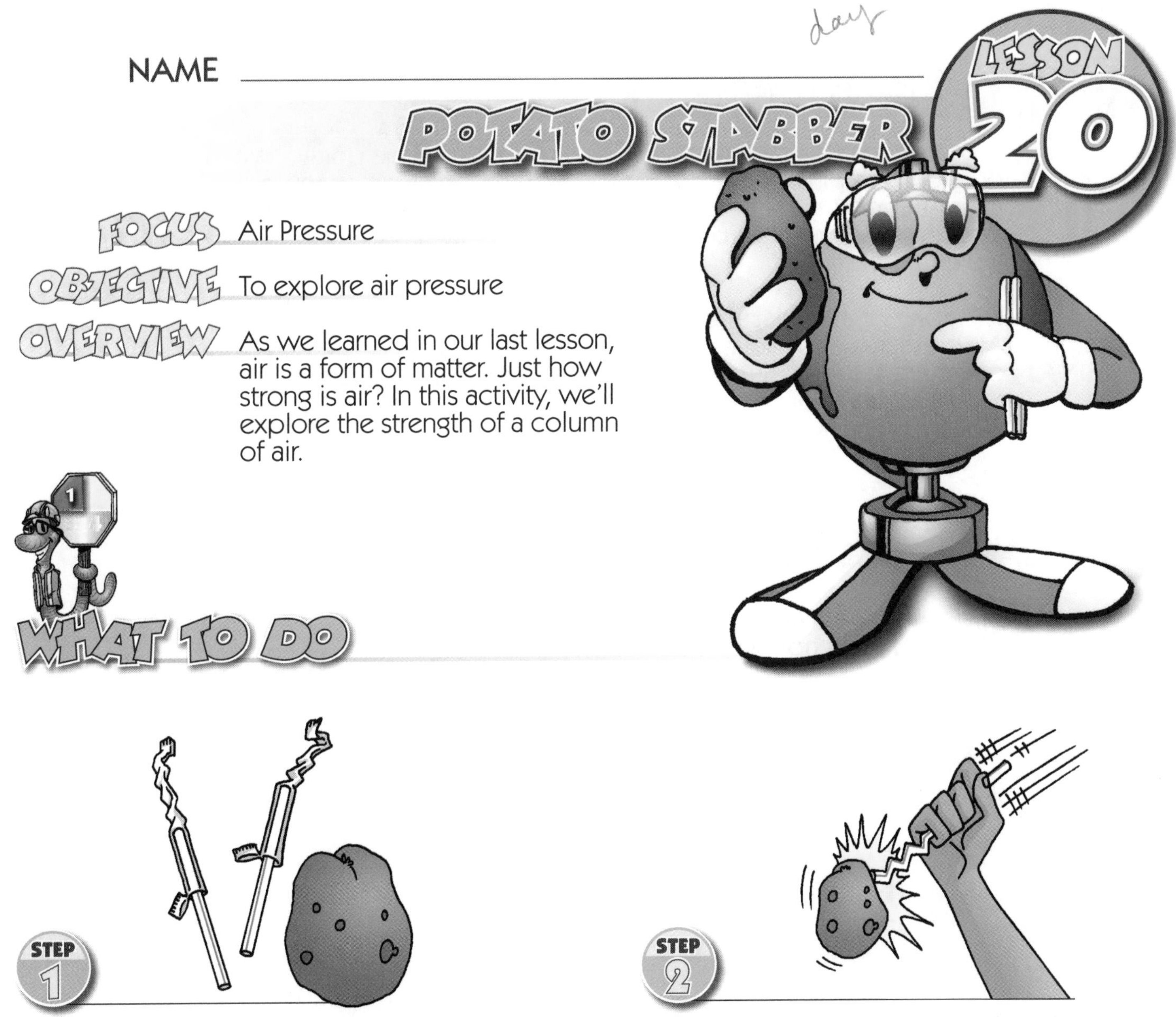

FOCUS Air Pressure

OBJECTIVE To explore air pressure

OVERVIEW As we learned in our last lesson, air is a form of matter. Just how strong is air? In this activity, we'll explore the strength of a column of air.

WHAT TO DO

STEP 1

Remove the paper covers from two soda straws and **place** the straws on your work surface. **Lay** a potato beside the straws. **Examine** these objects and **record** your observations on your journal page.

STEP 2

Hold the first straw as shown in the illustration. Try to **stab** it into the **potato**. Record the results.

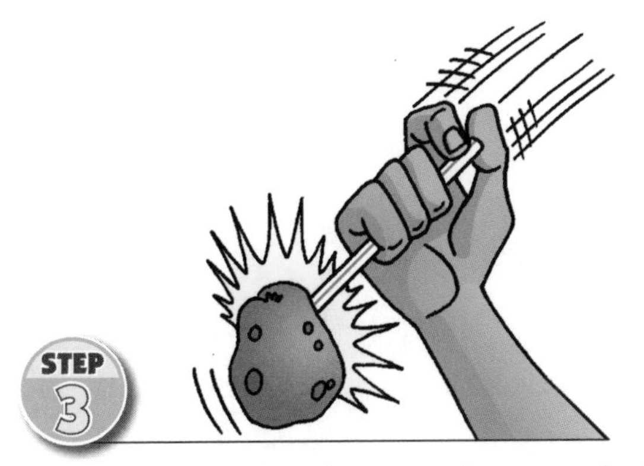

STEP 3

Repeat Step 2 using the second straw, only this time **seal** the end of the straw with your thumb (see illustration). **Hold** your thumb tight against the straw so no air can get out. **Record** the results.

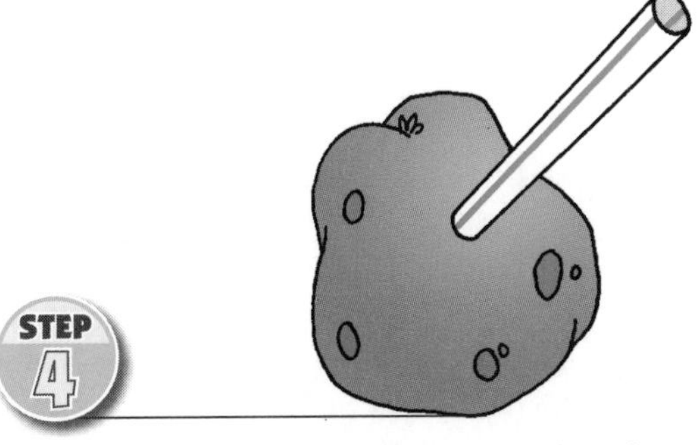

STEP 4

After everyone has had a turn, **examine** the "stabbed" potato. **Record** your observations. **Share** and **compare** observations with other research teams.

In Step 2, the straw wasn't strong enough to pierce the potato. When the straw hit the potato, the **air** inside the straw escaped and the straw crumpled. But in Step 3, your finger trapped a column of air inside the straw. When the straw hit the potato, the trapped air **compressed** and its **pressure** helped support the straw. The result was a pierced potato!

Examples of **air pressure** are all around us. The air pressure in tires holds up your car. Air pressure keeps your basketball from going flat. Air pressure is used to run many commercial tools. Air pressure in the **atmosphere** (the air you breathe) even affects the **weather**, creating highs or lows, sunny days or storms.

Is air a form of matter? Does it take up space? Explain how you know this.

Describe what happened when the straw hit the potato in Step 2. What caused this to happen?

Describe what happened when the straw hit the potato in Step 3. What caused this to happen?

Compare Step 2 with Step 3. How were they similar? How were they different?

Give at least three examples of air pressure and tell how they affect us.

Air has pressure. Air under pressure can do many things. Air pressure even affects the weather.

Isaiah 40:21 That weak little straw just didn't have the strength to pierce a tough potato. But when your finger trapped the air, it provided the strength needed to do what seemed impossible!

This Scripture reminds us that our strength comes from God. Your life may seem full of impossible tasks as you try to keep up with school work, the needs of your friends, and responsibilities at home. Just remember to spend time learning to trust God, waiting for him to guide you — and he will give you all the strength you need!

My Science Notes

NAME ______________________

WATERPROOF COTTON

LESSON 21

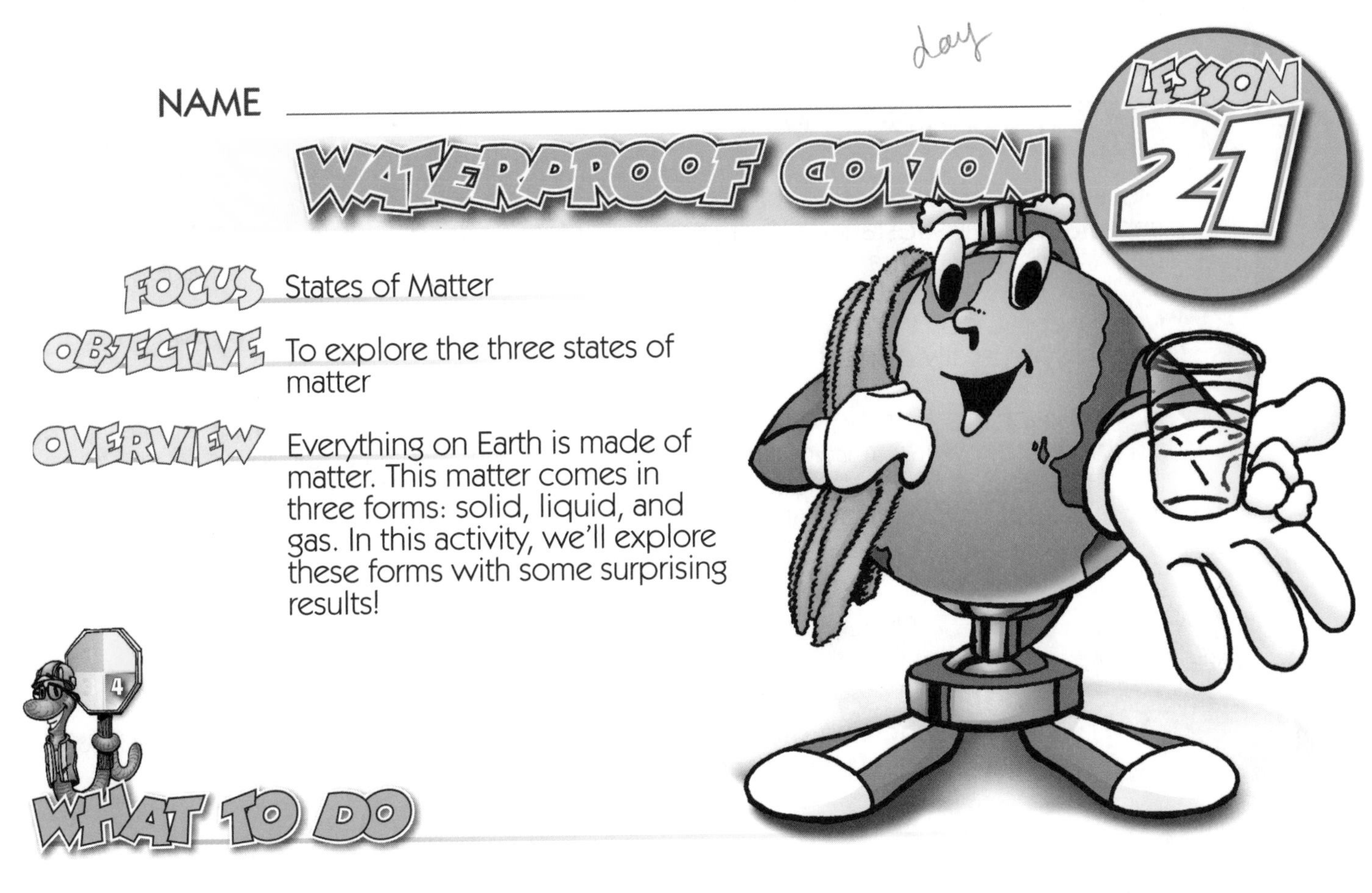

FOCUS States of Matter

OBJECTIVE To explore the three states of matter

OVERVIEW Everything on Earth is made of matter. This matter comes in three forms: solid, liquid, and gas. In this activity, we'll explore these forms with some surprising results!

WHAT TO DO

STEP 1

Hold a cotton ball above a cup of water. **Predict** what will happen to the cotton ball if you hold it under water. **Record** this prediction on your journal page.

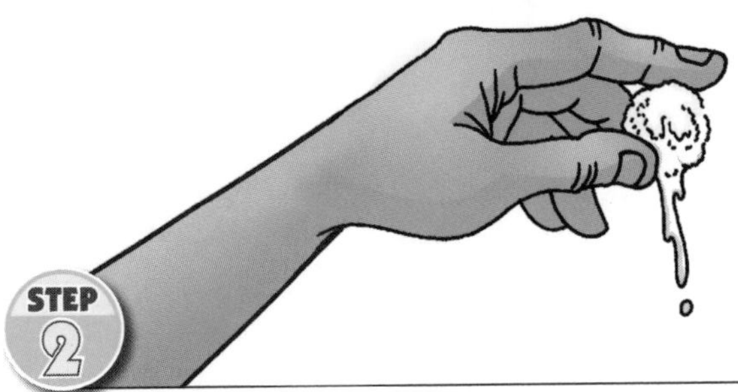

STEP 2

Push the cotton ball down into the cup. **Hold** it under the water for three seconds. Now **lift** it up and **hold** it above the cup. **Record** the results.

STEP 3

Watch as your teacher places cotton balls in a glass. **Predict** what will happen to the cotton balls when the glass is pushed under the water. **Record** this prediction on your journal page.

STEP 4

Watch closely as your teacher pushes the glass under the water. **Record** your observations. **Compare** the results with Step 2. **Share** and **compare** observations with your research team.

Most science books define **matter** as something that has **volume** (it takes up space) and **mass** (it has substance). There are three **states** (kinds) of matter: **solid**, **liquid**, and **gas**. In this activity, the cotton is a solid, the water is a liquid, and the air is a gas.

Many people don't understand how air and other gases can be matter. You can't see air and you usually don't think of air as having substance.

In Step 4 of this activity, your teacher trapped air inside a cup. When the cup was **pushed** under water, the volume and mass of the air pushed back on the water, keeping it out. Instead of water soaking the cotton ball (as in Step 2), the gas/matter/air in the cup kept the water out and the cotton ball dry!

What are the two main characteristics of matter? What does "volume" mean? What does "mass" mean?

What are the three states of matter? Give an example of each.

What did you predict in Step 1?
How did this prediction reflect what actually happened?

What did you predict in Step 3?
How did this prediction reflect what actually happened?

Based on what you've learned, explain why the cotton balls didn't get wet in Step 4.

Conclusion

Everything on Earth is made of matter. There are three states of matter: solid, liquid, and gas. Each of these states has unique characteristics that affect all of us and the world we live in.

Food for Thought

Exodus 14:16 Things don't always happen the way we think they will. In Step 3, you probably thought the cotton balls would get soaked, even with the protection of the cup. But that's not what happened at all!

Imagine the people of Israel facing the sea with Pharaoh's army closing in. They probably thought their only choice was to be killed or to drown! But God's mighty power parted the waters, and his people walked to safety on dry land. Whenever things get rough and you think you can't go any further, remember that amazing things are possible through God.

Journal My Science Notes

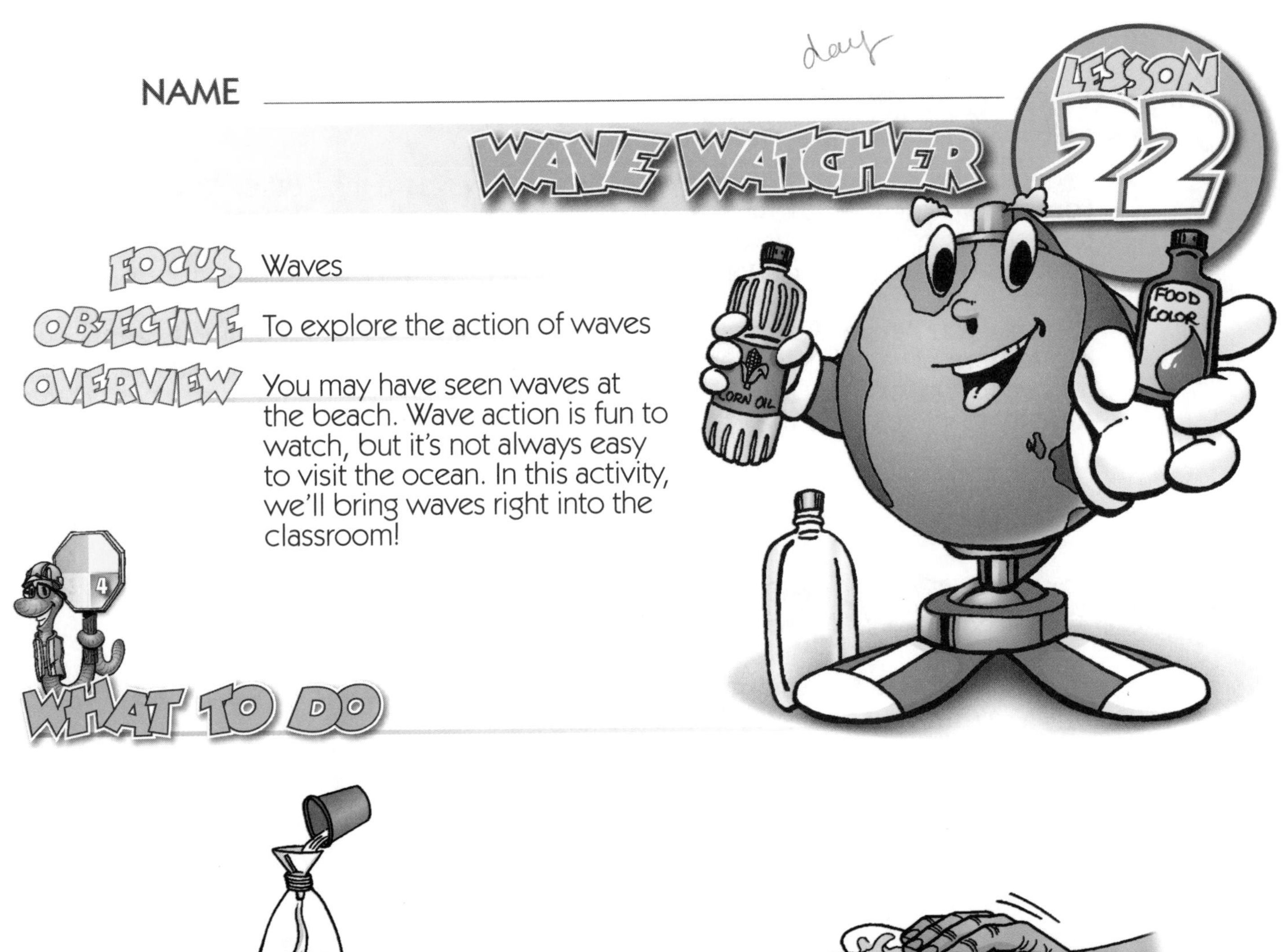

NAME ______________________

LESSON 22

WAVE WATCHER

FOCUS Waves

OBJECTIVE To explore the action of waves

OVERVIEW You may have seen waves at the beach. Wave action is fun to watch, but it's not always easy to visit the ocean. In this activity, we'll bring waves right into the classroom!

WHAT TO DO

STEP 1

Fill a bottle half full of water. **Add** a few drops of blue food coloring until the water looks like the ocean. Carefully **fasten** the lid back on the bottle.

STEP 2

Lay the bottle on its side. Slowly **rock** the bottle back and forth. Carefully **watch** the waves that form. **Make notes** in your journal about what you see.

STEP 3

Set the bottle upright and **remove** the lid. **Add** cooking oil to the water until the bottle is full. (Don't leave any air in the bottle.) **Fasten** the lid back on firmly.

STEP 4

Repeat Step 2. **Compare** what happens with what happened in Step 2. After everyone has had a turn, **rock** the bottle faster. **Share** and **compare** observations with your research team.

Your muscles provided the **energy** that created waves in the bottle. **Wind** (and other **forces**) provides the energy that creates waves in the ocean. **Waves** contain powerful amounts of energy. Waves pound shorelines, **eroding** the land. Waves rearrange beaches by moving sand. Waves carry **nutrients** needed for life around the planet.

To make your **Wave Watcher**, you used two **liquids** — oil and water. Even though the **volume** of each liquid is about the same, they formed two distinct **layers**. The oil moved to the top because it is less dense (lighter) than water. The water has greater **density** than the oil, so the water sank to the bottom.

What was the energy source for the waves in Step 2 and Step 4? What causes waves in the ocean?

What ingredients did you use to make your Wave Watcher? How were they similar? How were they different?

3. Compare the wave action in Step 2 with Step 4. How were they similar? How were they different?

4. Which has the greater density — water or oil? How do you know this?

5. Based on what you've learned, if you spilled cooking oil in a swimming pool would it float or sink? Why?

Waves are caused by wind and other forces. Waves help transfer energy around the Earth. Wave action can cause erosion. Wave energy can also move nutrients over vast distances.

James 1:5-8 The waves in your bottle didn't have much choice. You could make them move smoothly back and forth, or toss them around in confusion! They were instantly affected by whatever was happening around them.

This Scripture reminds us that our lives don't have to be jumbled and confused like endlessly tossed waves. When you learn to trust in God and listen to his voice, his love can fill your heart with peace — no matter what your surroundings!

My Science Notes

NAME ____________________

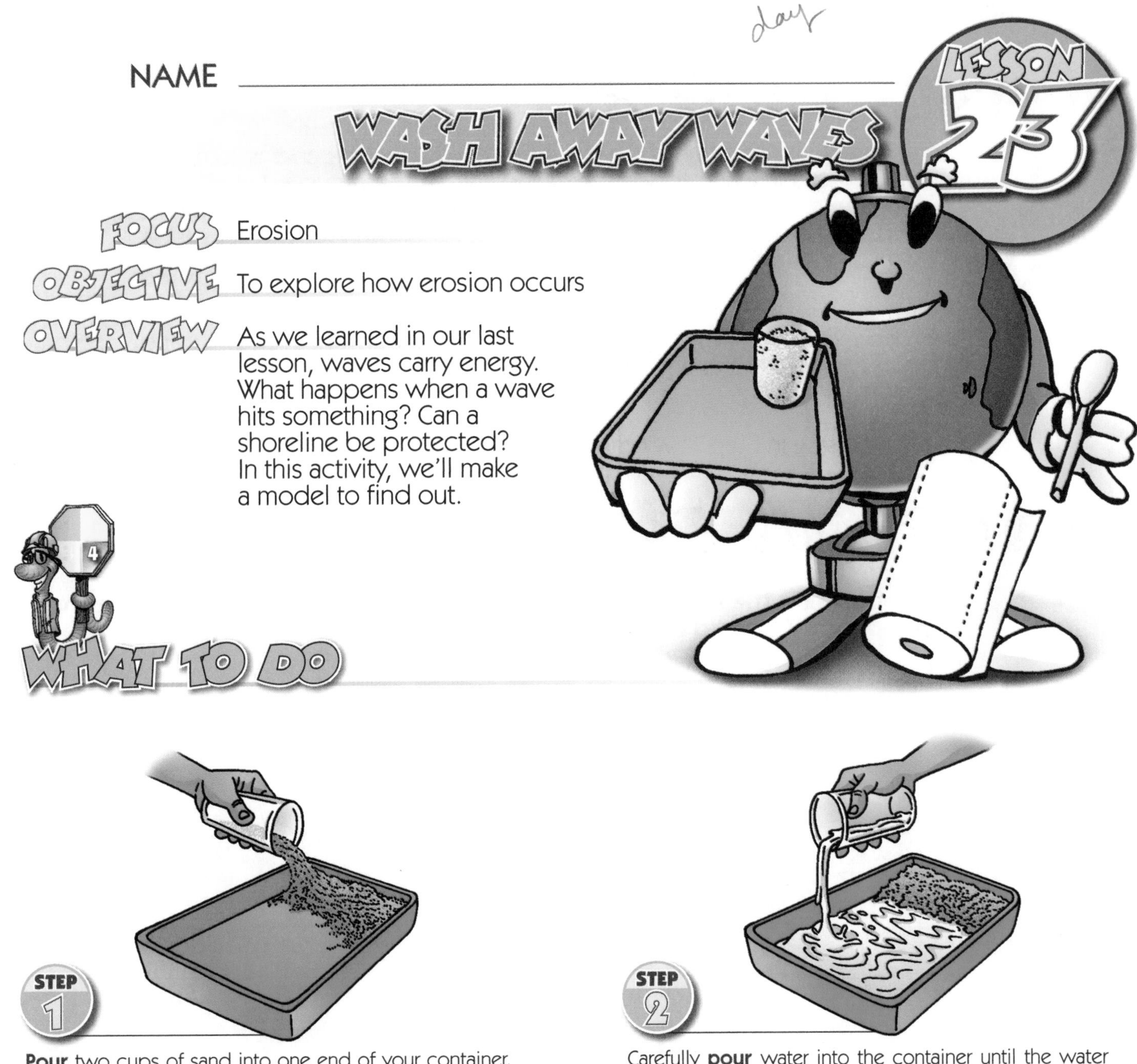

LESSON 23

WASH AWAY WAVES

FOCUS Erosion

OBJECTIVE To explore how erosion occurs

OVERVIEW As we learned in our last lesson, waves carry energy. What happens when a wave hits something? Can a shoreline be protected? In this activity, we'll make a model to find out.

WHAT TO DO

STEP 1

Pour two cups of sand into one end of your container. **Make** the pile as narrow and tall as you can. **Blow** gently on the sand. **Record** the results on your journal page.

STEP 2

Carefully **pour** water into the container until the water is a half inch deep. Don't let the water disturb the sand! **Observe** the sand and calm water. **Make notes** about what you see.

STEP 3

Use a plastic spoon to gently **splash** water toward the "shore" of the sand. **Repeat** this four times, observing what happens to the sand. **Record** the results.

STEP 4

Repeat Step 3, but **place** your hand in front of the sand. **Record** the results. **Share** and **compare** observations with your research team.

What Happened?

It takes **energy** to make anything move. You provided the energy to move the spoon, which moved the water toward the sand. When the moving water hit the sand, the sand began to move as well. Scientists call this shifting of materials by moving water **erosion**. Erosion can happen along a beach, the shore of a lake, a river bank, or even in the middle of a farmer's field after a heavy rain!

Erosion is a serious problem in some areas. To control erosion along the edges of huge lakes or ocean shorelines, engineers design special **barriers** to **absorb** or **divert** some of the moving water's energy. You **modeled** this in Step 4 when you used your hand as a barrier.

And as you saw in Step 1, wind erosion can also be a potential problem!

What We Learned

Describe the model you created in Step 1 and Step 2. What were the parts? What did each part represent?

Compare the "shore" in Step 2 with the same shore after Step 3. How were they similar? How were they different?

Describe what happened in Step 3.
When materials are shifted by moving water, what is it called?

4

Where did the energy come from in Step 3 and Step 4? What caused the shore to erode in Step 3? What kept it from eroding in Step 4?

5

Based on what you've learned, name at least one thing that determines the amount of erosion.

Erosion occurs due to the energy in moving water. Unless this energy is absorbed or diverted by a barrier, erosion can damage a shoreline.

Matthew 17:20-21 Erosion happens rapidly on soft, sandy surfaces. It takes a little longer when the ground is tougher. But given enough time, erosion can even wear away solid rock!

This Scripture reminds us that our lives are only secure when they're built on the "rock" of a relationship with God. Take time every day to talk to God. Look for ways to learn more about his love and power. Then when problems surround you, like angry waves crashing in, you'll be safe and secure in God's care.

My Science Notes

NAME ______________________________

LESSON 24
FLOOD IN A JAR

FOCUS Sediments

OBJECTIVE To explore how water separates materials

OVERVIEW You've probably seen pictures of the enormous damage floods can do. Usually there's mud everywhere! Where does all that mud come from? In this activity, we'll make a model to find out.

WHAT TO DO

STEP 1

Add equal amounts of sand, potting soil, and aquarium gravel to a glass jar. **Rotate** the jar slowly and **look** at the contents. **Make notes** on your journal page about what you see.

STEP 2

Slowly **add** water until the water level is slightly higher than the materials in the jar. **Fasten** the lid firmly, then slowly **rotate** the jar and **look** at the contents. **Record** your observations.

STEP 3

Shake the jar for about 15 seconds. Quickly **set** it on your work surface and **observe** what's happening inside. (Don't touch the jar again until the activity is completed.) **Record** your observations.

STEP 4

Wait three minutes. **Look** at the contents of the jar again. **Record** what you see. **Share** and **compare** observations with your research team.

When flood waters hit, there's usually a huge amount of material being carried along in the water. The **force** of the water holds and carries these materials until the water slows down. Scientists call this a **suspension**. You demonstrated a suspension when you shook the jar. As long as you kept shaking, the heavy pieces (like gravel) were just mixed in with all the rest.

But when you stopped shaking, **gravity** began to **pull** things down. As water slows, it loses **energy**. Since it takes more energy to keep large materials suspended, they fall first. And since it takes less energy to keep light materials suspended, they fall last and end up on top. This process **sorts** the materials into **layers**, and these layers are known as **sediments** (materials that are deposited by wind or water action).

What ingredients did you add to the jar in Step 1?
What happened to these ingredients as you rotated the jar?

What ingredient did you add in Step 2?
What affect did this have on the materials already in the jar?

Describe what happened in Step 3.
What did the force of the moving water do to the heavy pieces?

4

Describe the contents of the jar after Step 4. Describe the top layer. How is it different from the other layers?

5

A flood hits your school, sweeping away a stack of test papers, a pile of marbles, and a large iron doorstop. It dumps the debris in a nearby field. Based on what you've learned, where would you look for each item?

CONCLUSION

Moving water has energy that can carry materials long distances in a mixture called a suspension. As moving water slows, the heaviest materials settle out first. The lightest materials can form sediments.

Psalm 29:10 Your model demonstrated how the power of moving water can suspend heavy objects. You may have seen pictures of floods sweeping away bridges, huge trees, even homes and businesses! Moving water has tremendous power!

Although a raging flood may seem like one of the most powerful forces on Earth, this Scripture reminds us that even during the largest flood ever known, God was still in control. What tremendous power! Isn't it amazing to think that this powerful God knows you individually and that his love for you will last forever?

My Science Notes

NAME ______________________________

LESSON 25

SKY IN A JAR

FOCUS Earth's Atmosphere

OBJECTIVE To explore how light scatters

OVERVIEW Sunlight usually appears as some shade of yellow or orange. But then why does sunlight make the sky look blue? In this activity, we'll create a model to help us find out.

WHAT TO DO

Sit quietly as your teacher darkens the room. Now **turn on** your flashlight and **shine** it at the top of a wall. **Observe** the flashlight beam. **Make notes** on your journal page about what you see.

Fill a jar or glass with water. **Shine** the flashlight through the water. **Compare** the flashlight's beam with how it looked in Step 1. **Record** your observations.

Keep shining the flashlight through the water. Using a pipette, **add** one drop of milk to the water. **Observe** the result and **record** it in your journal.

Repeat Step 3, but **stir** the water gently as you add the milk. **Add** several more drops, observing changes in the flashlight's beam. **Record** the results. **Share** and **compare** observations with your research team.

What Happened?

In Step 2, you shined **light** through air, then glass, then water, and out the other side. The light went through without much trouble since all of these materials are **transparent**. Then you began to add milk to the water. The protein and fat **molecules** in the milk are much larger than water molecules. When you shined the light again, the beam hit these large particles and began bouncing off, creating the changes you saw.

Scientists call this effect **scattering**. Our model helped us see it on a small scale. Did you notice that the light scattered by your model was a light bluish gray? This is because blue light scatters easily due to its wavelength. On a much larger scale, light from the sun shines though space, then hits the air molecules in Earth's atmosphere. The sunlight scatters, and we get a blue sky!

What We Learned

1. What did the flashlight beam represent in this model? What did the molecules in the milk represent?

2. Why was there no significant light change in Step 2? What did the materials involved have in common?

3 Describe why the light began to change in Step 3.
What do scientists call this effect?

4 Compare the light in Step 1 with the light in Step 4.
How were they similar? How were they different?

Based on what you've learned, explain why the sky looks blue even though sunlight appears to be yellow or orange.

Sunlight scatters when it strikes air molecules in Earth's atmosphere. Since blue light scatters the easiest, it causes Earth's sky to look blue.

I Corinthians 10:31 - 11:1 This model helped us see how light can be scattered. Using a model is certainly much easier than taking a trip into space! A model is often a great tool for helping us understand difficult ideas in an easy way.

Jesus is the ultimate model for our lives. Paul reminds us that rather than doing what we like best, we should focus on doing what is best for others. We should always follow the example Jesus set for us. The more time we spend with God, the better we will understand how important this is, and the easier it will become.

My Science Notes

NAME ______________________

Week

LESSON 26
FAST FOSSIL

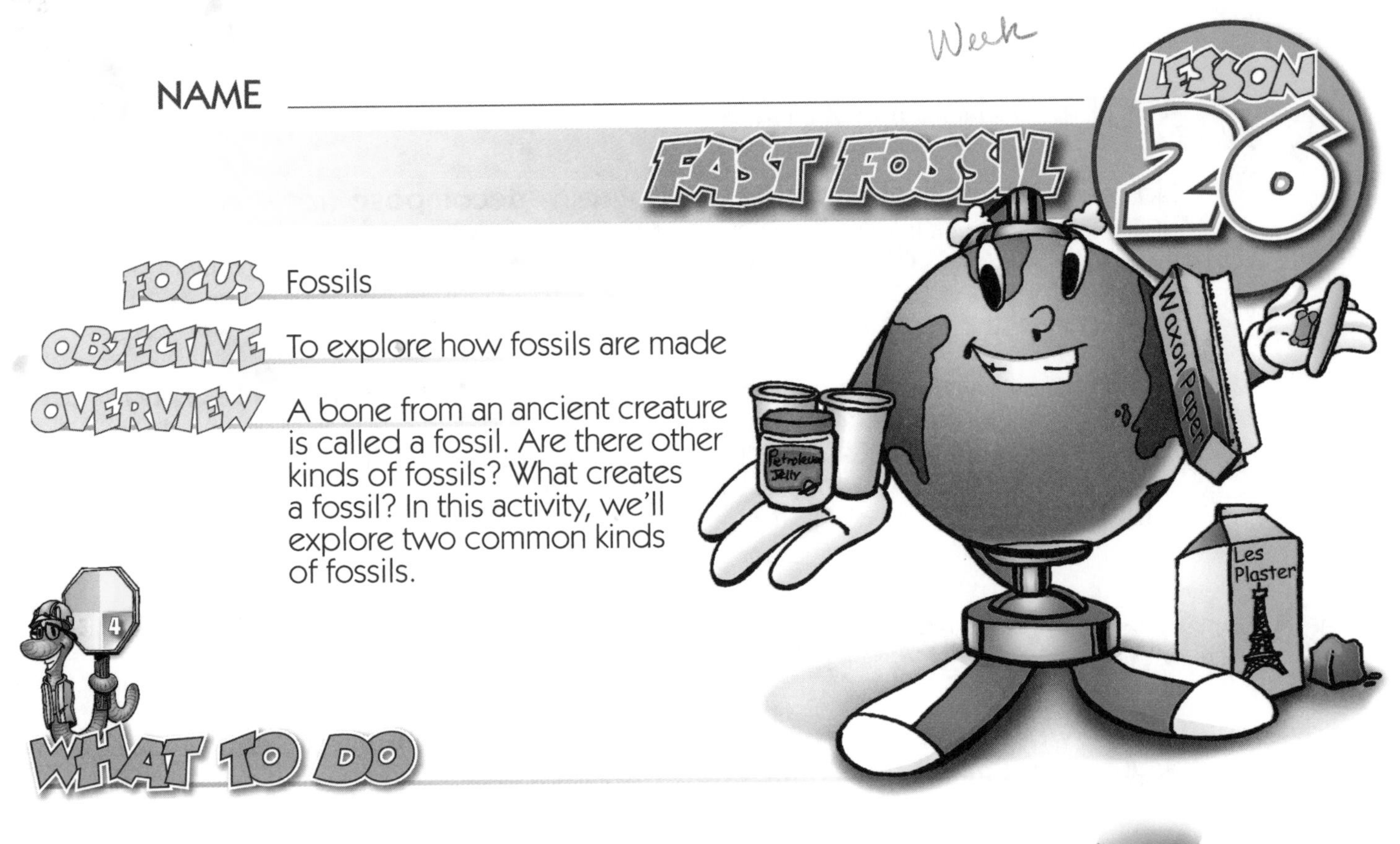

FOCUS Fossils

OBJECTIVE To explore how fossils are made

OVERVIEW A bone from an ancient creature is called a fossil. Are there other kinds of fossils? What creates a fossil? In this activity, we'll explore two common kinds of fossils.

WHAT TO DO

STEP 1

Carefully **examine** the "bone" that will become your fossil. **Make notes** about its size, shape, and general appearance. Now **cover** one side of the bone with a thin, even coat of petroleum jelly.

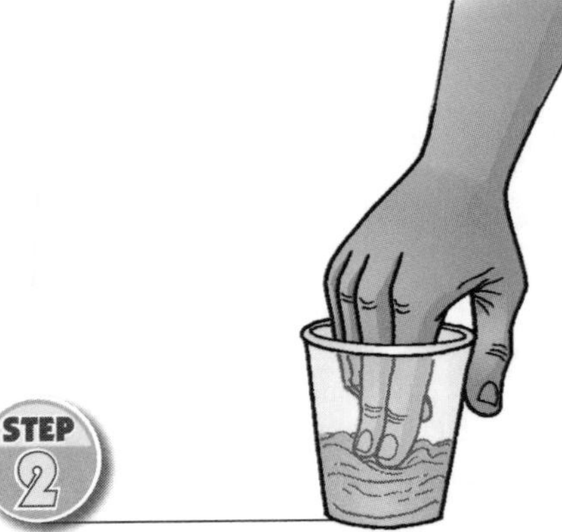

STEP 2

Mold a lump of clay into a smooth layer in the bottom of a paper cup. (The layer must be thicker than the bone you're using.) **Cover** the clay with a thin, even coat of petroleum jelly.

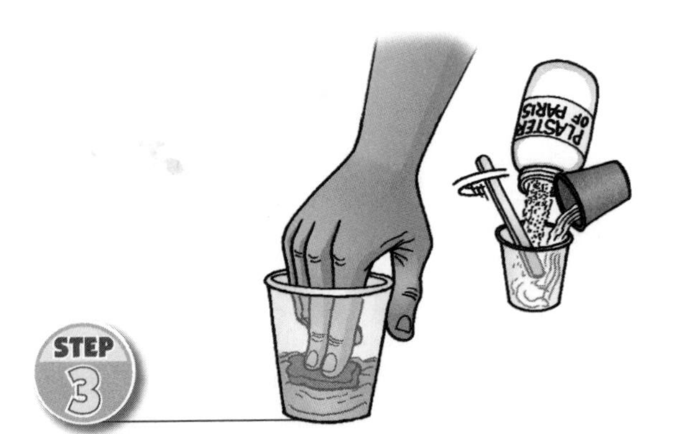

STEP 3

Push the coated side of the bone into the clay, then carefully remove it. In a second cup, **mix up** plaster of Paris. (Your teacher will explain how to do this.) Carefully **pour** the plaster into the first cup.

STEP 4

[next day] After letting the plaster harden overnight, **tear** away the sides of the paper cup. Gently **pry** the clay off the plaster. **Examine** the results. **Share** and **compare** observations with your research team.

Normally when living things **die**, their remains **decompose** (rot away) quickly. But under special conditions, the object or imprint is **preserved**, sometimes even turning to stone. Scientists define a **fossil** as any preserved part of an ancient living thing, or evidence that it once existed. For instance, a footprint is a kind of fossil even though every part of the creature that made it is long gone!

In this activity, you made **models** of two types of fossils: a **mold fossil** and a **cast fossil**. The dent you made in the clay (using the bone) created a mold. An imprint like this is called a mold fossil. The plaster you poured in the mold hardened to form a cast. A cast fossil looks a lot like the original object. Notice that neither of these fossils is the object itself, just a sign that it once existed.

What was the purpose of the petroleum jelly in Step 1 and Step 2? What might have happened without it?

How were mold fossils made? Give an example of a mold fossil.

3 How were cast fossils made? Which looks more like the original, a cast fossil or mold fossil? Why?

4 Which was more detailed, the original object or the cast fossil model you made? What does this tell you about the study of fossils?

5 Based on what you've learned, name two things you could know about an object from a fossil. Name two things you could not know about an object from a fossil.

A fossil is any preserved part of an ancient living thing, or evidence that it once existed. Although there are many missing details, fossils offer clues about ancient plants and animals.

John 14:9 In this activity, you learned how different types of fossils are made, and how they offer clues about ancient plants and animals. But at best, fossils only give us a fuzzy picture of the past. It's like putting a puzzle together with some of the pieces missing!

Fortunately, God is not like that! Through stories, lessons, and parables, the Scriptures give us a clear picture of God's character. Jesus' life was the ultimate reflection of God's great love. God is willing to help you know him better and better if you'll only learn to trust in him.

My Science Notes

NAME ______________________ Week

LESSON 27

STAR SEARCH

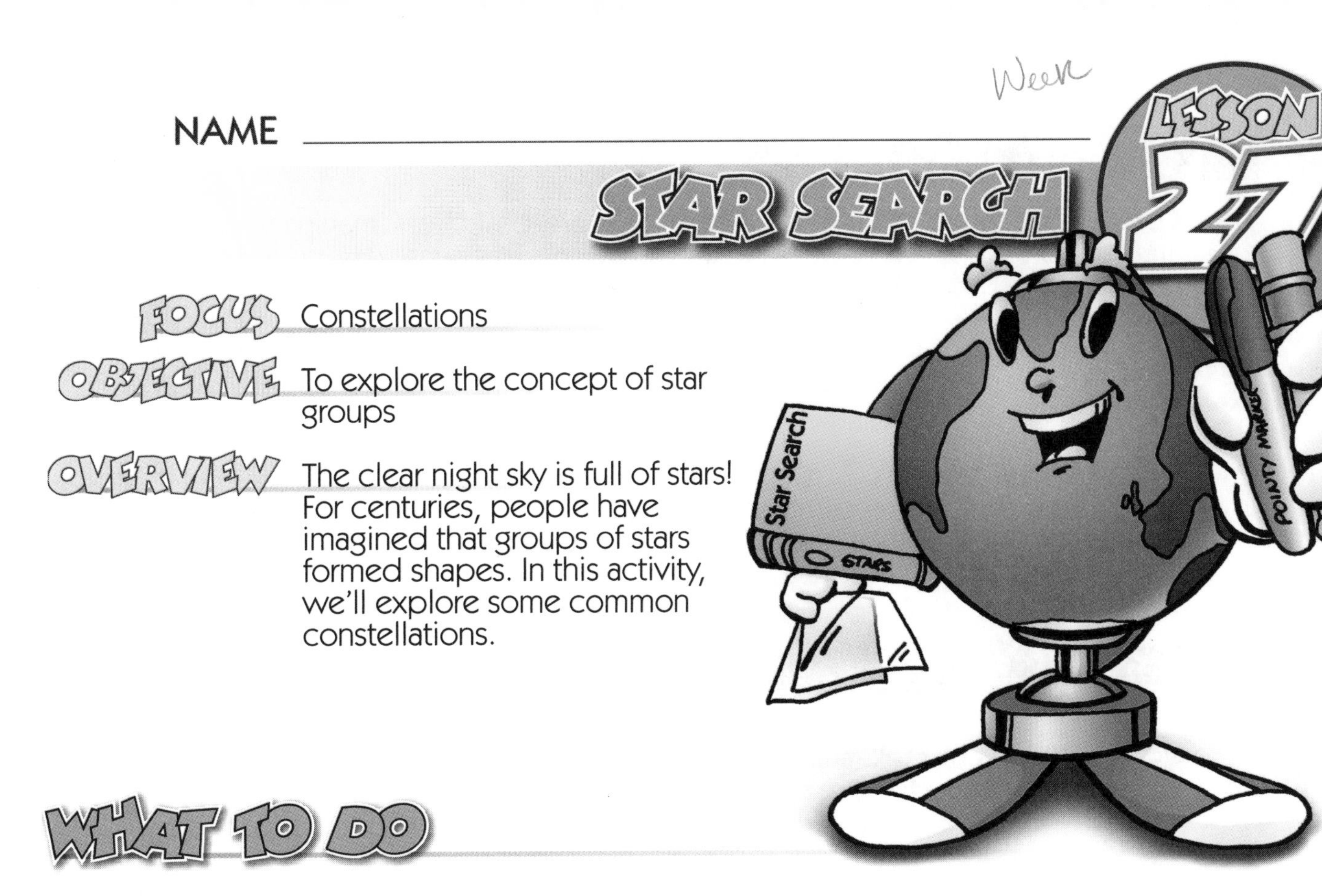

FOCUS Constellations

OBJECTIVE To explore the concept of star groups

OVERVIEW The clear night sky is full of stars! For centuries, people have imagined that groups of stars formed shapes. In this activity, we'll explore some common constellations.

WHAT TO DO

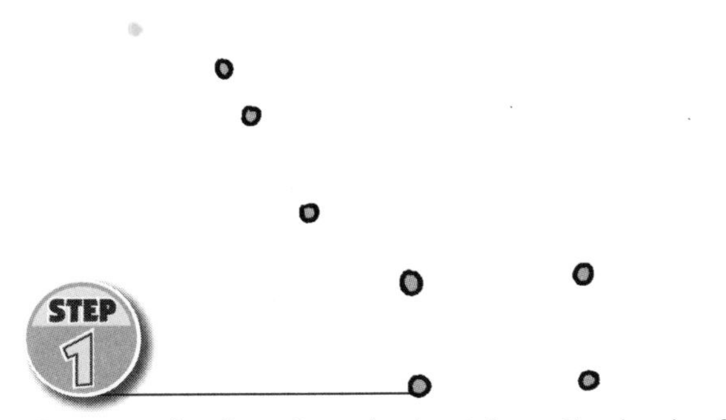

STEP 1

Remove the **Star Search** sheet from the back of your worktext (p. 165). **Choose** a constellation, or use the one assigned by your teacher. **Count** the stars in this group and **make notes** in your journal about the arrangement of these stars.

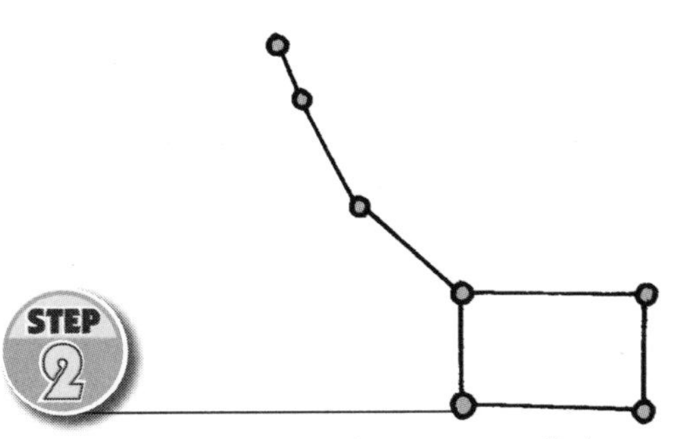

STEP 2

Draw a larger version of your constellation on a sheet of paper. **Connect** the stars as shown on your sheet. Use the Internet or an encyclopedia to **research** your constellation. **Record** your findings.

STEP 3

Choose the best drawing from your team. **Cover** it with a transparent sheet of plastic. Using a black marker, **draw** a large dot for each star. **Label** any individual stars your team found in their research.

STEP 4

Choose a team member to present your findings to the class. **Ask** them to share any interesting information your team found. **Listen** closely to other teams' presentations and **make notes** in your journal.

What Happened?

Constellations are groupings of **stars** that are based on imaginary shapes people created to help them remember the night sky. This was very important to ancient people who used stars to navigate the oceans or to determine directions after dark on a long journey!

There are 88 constellations that can be seen from the Earth's surface. Many of the names and arrangements go back to ancient times.

But since **sky patterns** shift over time, the constellations you see are not quite the same as the ones that ancients saw. In fact, if you were far out in space, the arrangements wouldn't look the same at all! Still, constellations are a fun way to memorize star positions and to enjoy the night sky.

What We Learned

1. Which constellation did your research team study? How many stars did it include? Describe their arrangement.

2. What did you discover about your constellation in Step 2? Did any stars in your constellation have names? If so, what were they?

Name the constellations studied by other research teams.
Describe at least one thing about each.

Why did ancient people give star groupings names?
What practical purpose did it serve?

Based on what you've learned, would the constellations look the same from a million miles out in space? Why or why not?

Constellations are star groups based on imaginary shapes that ancient people created to help them remember the night sky. The shape of a constellation changes with time and distance.

Matthew 2:1-2 This activity has given you a glimpse into the wonders of the night sky. Did you know that there are over 9,000 stars visible without using a telescope? Millions more can be seen with the powerful tools scientists have developed. And the deeper into space they look, the more stars they find!

As ancient astronomers studied the night sky, they saw something surprising — a new star! That wondrous light led them to the King of the universe, the tiny child Jesus. Why not follow their example and let Jesus into your heart? Be a "star" and let the love of God shine though you!

My Science Notes

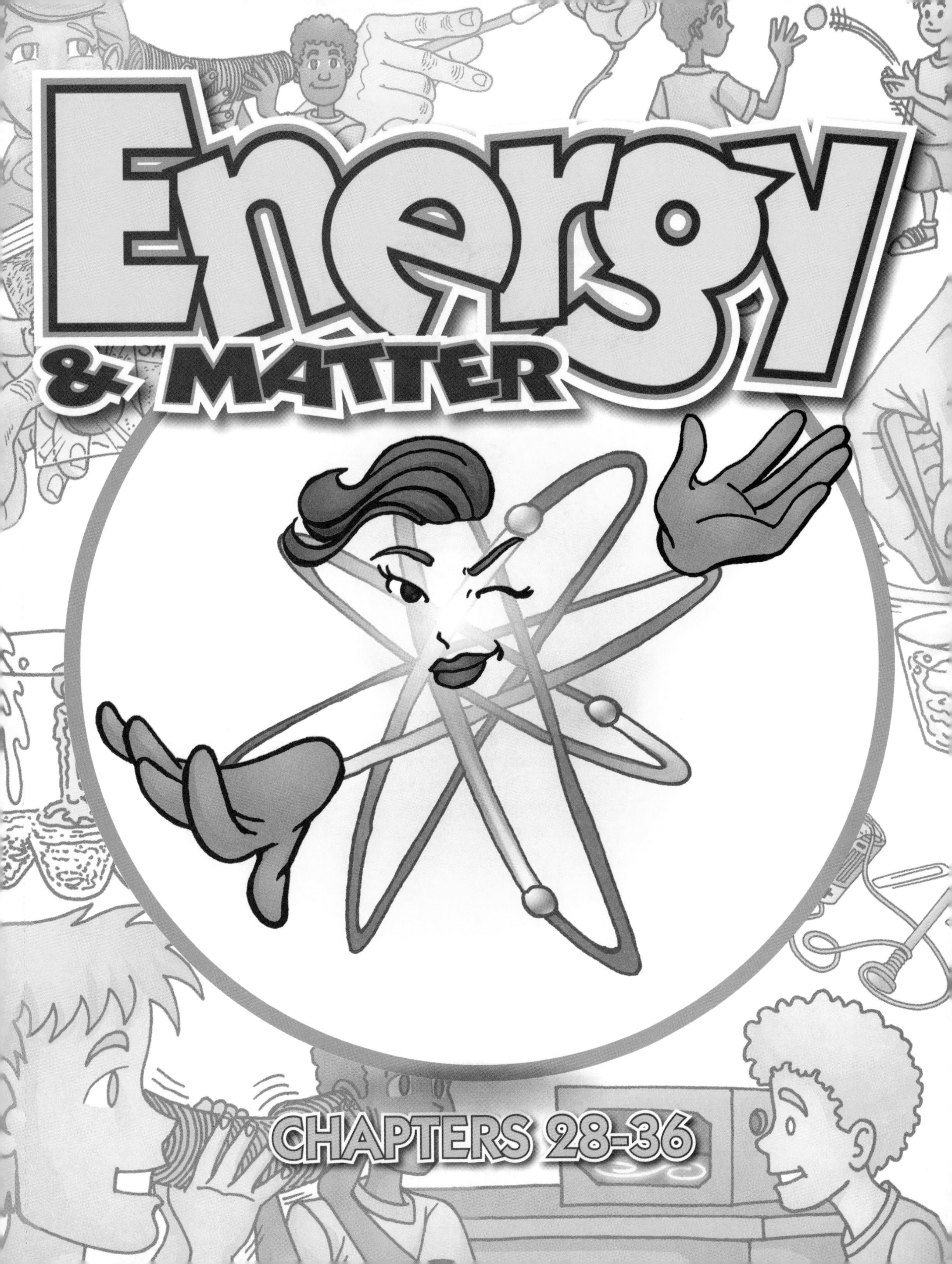
Energy
& MATTER
CHAPTERS 28-36

Energy and Matter

In this section, you'll learn about the different **states** and unique **properties** of **matter**. You'll discover new things about light and sound. You'll explore physical and chemical reactions. Some grades will explore related concepts like circuits, currents, and convection.

NAME ______________________________

LESSON 28
WAVE MAKER

FOCUS Transfer of Energy

OBJECTIVE To explore how energy moves as waves

OVERVIEW You see sunlight on a leaf. You hear someone talking on the phone. You cook a pizza in the oven. Did you know all these activities are related? In this activity, we'll find out how!

WHAT TO DO

Choose a member of your research team. **Stand** facing each other. Now each of you **pick up** one end of the slinky. **Back** away from each other until the slinky is stretched nearly level. (**Hold** on tight!)

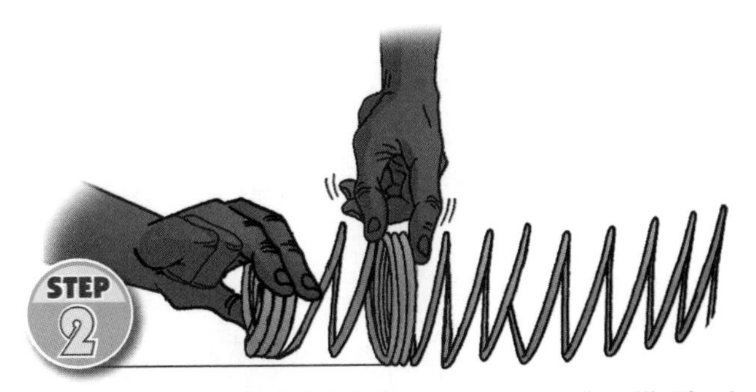

Ask you partner to hold their end completely still. **Pinch** together five rings of the slinky near where you're holding it. On the count of three, **release** the rings. **Record** the results.

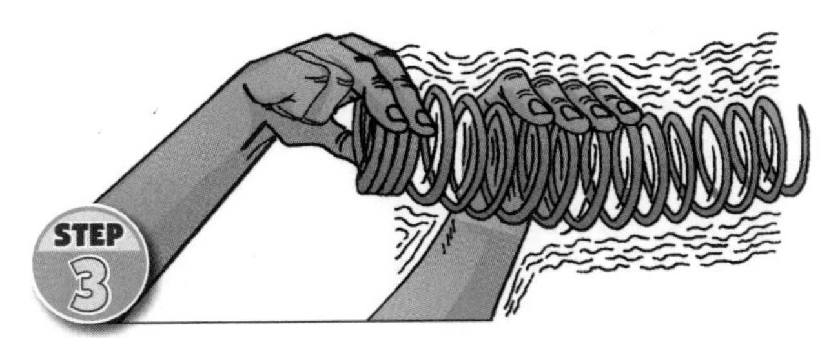

Hold your end of the slinky completely still. **Ask** your partner to wiggle their end back and forth two times, then **stop**. **Record** the results.

Hold the slinky completely still. On three, both of you **wiggle** your end one time to the right and back, then **stop**. **Record** the results. Once everyone on your team has had a turn, **share** and **compare** observations.

What Happened?

It takes **energy** to make anything move. Your muscles provided the energy for this activity. When you wiggled your end of the slinky, the energy produced a **wave**. The wave traveled down the slinky to your partner, then it bounced back. Scientists call this **reflection**. (Drop a pebble in a bucket of water and you can see reflection when the wave bounces back from the sides.)

Many forms of energy travel in waves. You see the sunlight on a leaf — that **light** is traveling in waves. You hear someone talking on the phone — the **sound** is traveling in waves. You cook a pizza in the oven — the **heat** is traveling in waves.

Although we rarely think about it, many of our daily activities involve energy traveling in waves.

What We Learned

In Step 1, what is connecting the two people? Why isn't the slinky moving?

Describe what happened in Step 2. What was transferred from one end of the slinky to the other? What form did it take as it traveled?

3 Compare Step 2 with Step 3. How were they similar? How were they different?

4 Name at least three forms of energy that travel in waves. Give an example of each.

5 What might happen to the wave if a third person held the slinky in the middle? Compare this to the erosion barrier in Lesson 23?

CONCLUSION

It takes energy to make anything move. Some forms of energy travel in waves (light, heat, sound, etc.). Waves can move in different ways.

FOOD FOR THOUGHT

1 Thessalonians 5:16-18 In this activity, you moved the slinky to create a wave, which traveled down the slinky to your partner, then bounced back.

Prayer can also be like that. When you pray, your prayer travels to God, then an answer bounces back. God always answers prayers, even though the answer is sometimes "no" or "wait a while." God always wants what is best for you. Keep on praying, learn to trust him, and be thankful for the blessings he provides.

JOURNAL My Science Notes

NAME ______________________ day

MILK JUG MEGAPHONE

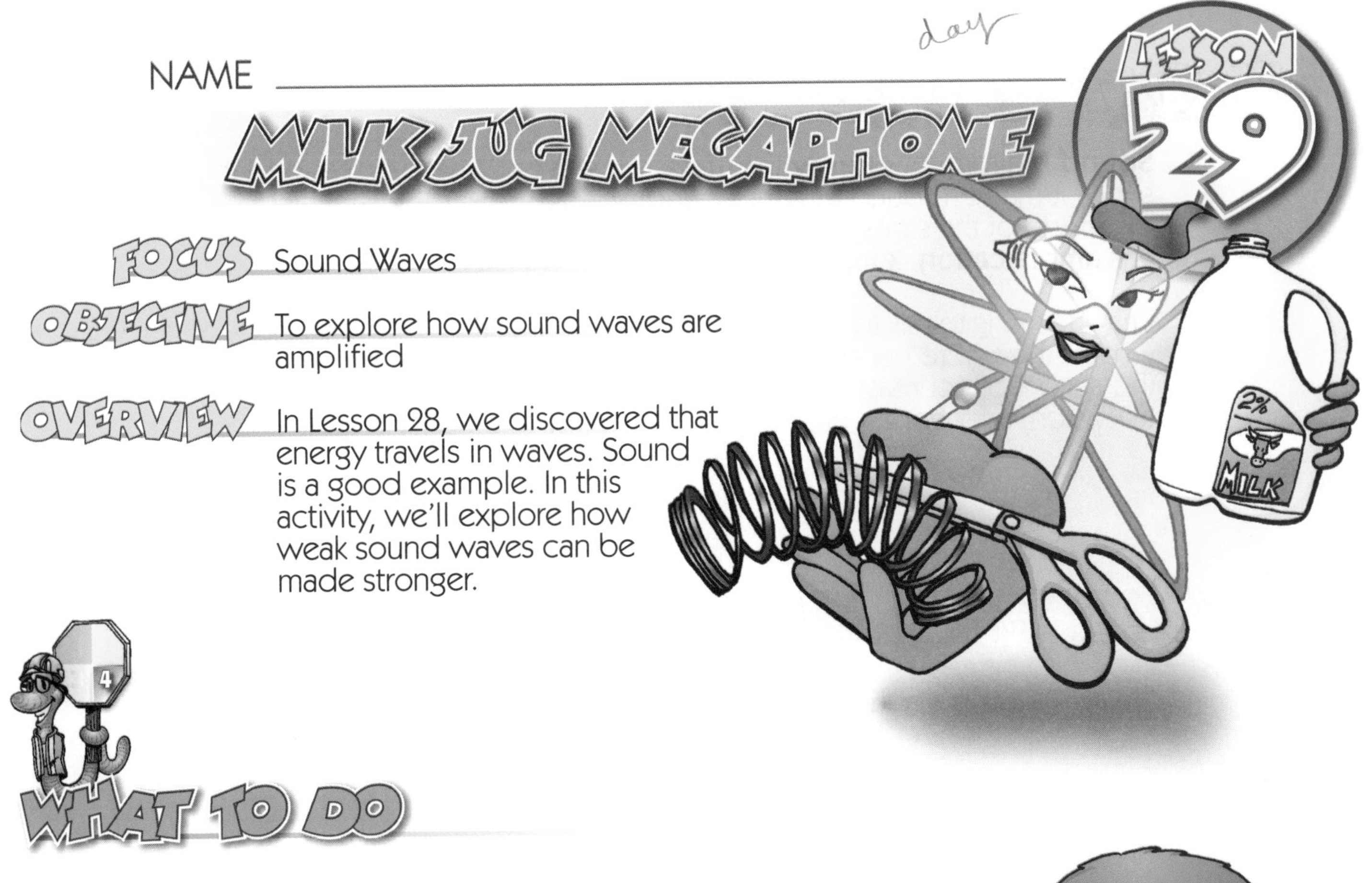

FOCUS Sound Waves

OBJECTIVE To explore how sound waves are amplified

OVERVIEW In Lesson 28, we discovered that energy travels in waves. Sound is a good example. In this activity, we'll explore how weak sound waves can be made stronger.

WHAT TO DO

Choose a member of your research team. **Stand** facing each other. Now each of you **pick up** one end of the slinky. **Back** away from each other until the slinky is stretched nearly level. (**Hold** on tight!)

Carefully **hold** the slinky up to your ear. **Ask** your partner to softly tap the other end of the slinky with a pencil. **Record** the results. **Trade** places and **repeat** this step.

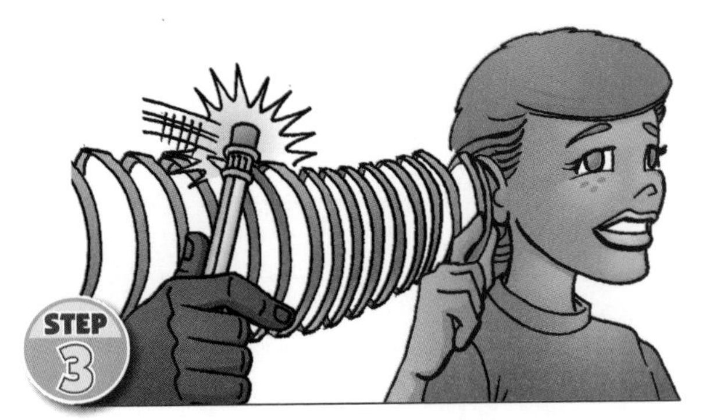

Repeat Step 2, but this time **tap** the pencil harder. **Record** the results. Now **cut** the bottom off a milk jug. **Cut** through the handle, then **slip** two end rings of the slinky onto the handle. **Tape** to secure.

Aim the open end of the jug at your ear. **Repeat** the taps from Step 2 and Step 3. **Record** the results. **Trade** places and **repeat** this step. Once everyone on your team has had a turn, **share** and **compare** observations.

What Happened?

Sound is a form of **energy** that travels in **waves**. When you tapped the pencil on the slinky, it made a sound. But the sound was very weak. By attaching the milk jug to the other end of the slinky, you created a device to make the sound louder. Scientists call this **amplification**. Amplification can help us hear or see weak waves.

Imagine trying to talk to someone across a football field. A megaphone might help, but what if they were hundreds of miles away? A telephone provides a way to **transfer** the sound of your voice over great distances. Most communication devices (radio, television, telephones, etc.) rely on this ability to **amplify** energy waves.

What We Learned

1. What is sound a form of? How does it travel?

2. Describe what happened in Step 2 and Step 3. How difficult was it to hear the pencil tapping?

3 Compare Step 2 with Step 4. How were they similar? How were they different?

4 What was the purpose of the milk jug in Step 4?
What is the process of making weak waves stronger called?

5 Based on what you've learned, list at least three devices that use amplification and explain their purpose.

Conclusion

Sound travels in waves. Weak waves can be amplified. Amplified waves can transfer energy over great distances and make weak waves easier to hear or see.

Food for Thought

Romans 8:38-39 The pencil taps were solid, but they weren't very loud. The milk jug amplified that sound so it was easier to hear. Now imagine placing a microphone inside the milk jug, and hooking it up to a loudspeaker. A simple pencil tap might sound like an explosion!

Most of us have family and friends who love us. That's a good feeling, isn't it? But that love is only a "weak wave" compared to God's love. This Scripture tells us that nothing can separate us from God's love — not angels or devils or fear or even death itself! Spend time with God each day, and experience the awesome power of his love.

Journal **My Science Notes**

NAME ____

LESSON 30

SUNSHINE STRING

FOCUS Energy Conversion

OBJECTIVE To explore how energy changes form

OVERVIEW Energy comes in many forms. Sometimes it's helpful to convert energy from one form to another. In this activity, we'll explore energy conversion with a special piece of string!

WHAT TO DO

Remove the string from your materials kit. **Examine** the string closely. **Record** your observations in your journal. Now **place** the string in a dark place overnight.

[next day] **Cut** a hole in one end of a shoebox. **Remove** the string from the dark place, quickly **place** it in the shoebox, and **replace** the lid. **Look** through the hole at the string. **Record** your observations.

Remove the string from the shoebox. **Place** it in strong sunlight (your teacher can suggest a spot). **Leave** the string in the light for at least 10 minutes.

Place the string back in the box. **Look** through the hole at the string. **Record** what you see. **Examine** the string carefully for any other changes. **Share** and **compare** observations with other research teams.

This string contains a special material that **absorbs** and **releases** light **energy**. Think of it as a kind of sponge that soaks up **light**, then gives it back. Scientists call such materials **phosphorescent** (fos' fo res' ent). Your phosphorescent string stored the sunlight in Step 3, then released the stored light in Step 4.

Changing the **form** of **energy** is very important in our modern world. The light, heat, air conditioning, and electricity in your school all involve energy changing forms. Without this ability to **change** and **store** energy our way of life would be vastly different!

Why was it important to start this activity by placing the string in the dark?

Describe Step 2. What did the string look like in the box?

Why was it important to place the string in direct sunlight? What did this allow the string to do?

Compare Step 2 with Step 4. How were they similar? How were they different?

Why is the ability to change or store energy important in our modern world?

Conclusion

Energy can change forms. A change in forms often involves storing energy for later use. Materials that store and release light energy are called phosphorescent.

Exodus 34:28, 29 When your Sunshine String was left alone in the dark, it became rather dull and uninteresting. But when it was exposed to the sun, it developed a pleasing glow!

This Scripture talks about a special time Moses spent with God. When he returned to the people, his face glowed brightly because he had seen God face-to-face. Without God's love in our hearts, we're often dull and uninteresting inside. But when our hearts are filled with God's love, we overflow with love and light!

My Science Notes

NAME

Lesson 31

Bug's Eye View

FOCUS Images

OBJECTIVE To explore how a lens affects an image

OVERVIEW Different creatures have different types of eyes with different lenses. Ever wonder what the world looks like through a bug's eye? In this activity, we'll use a special lens to find out!

Choose a member of your research team. **Stand** facing each other about five feet apart. Gently **toss** a ball to your partner so they can catch it. **Ask** them to toss it back. **Record** the results in your journal.

Repeat Step 1, but this time make the tossing and catching a little tougher by closing one eye. **Record** the results in your journal.

Standing in the same position, **close** one eye. Now with your open eye, **look** at your partner through the "bug's eye." **Record** the results. Let your partner try the bug's eye. **Discuss** your observations.

Repeat Step 1, but use only the bug's eye to see. **Record** the results. Make sure all members of your research team get a turn. **Share** and **compare** observations.

A **lens** focuses **light** to produce an **image**. Many devices use lenses (microscope, telescope, camera, etc.) to help us see differently. The device we used is a **kaleido-scope**. It combines several lenses, resulting in multiple images. Look at your partner with and without the kaleidoscope to see the difference between one lens (the one in your eye) and many (the ones in the kaleidoscope).

Some **insects** like flies have eyes with multiple lenses. Scientists call this arrangement a **compound eye**. Compound eyes work well for a fly, but not for you since you don't have the **brain** of a fly! When you tried to catch the ball using the compound eye, your human brain couldn't decide what to do with all those images. This made it nearly impossible to catch the ball.

What was the name of the device used in this activity? What makes it produce multiple images?

Describe what your partner looked like in Step 3. What created this effect?

3 Describe Step 4. What made it hard to catch the ball?
Why was this a problem?

4 What is an eye with multiple lenses called?
What type of creature has such eyes?

Name at least two additional devices that use a lens to produce an image. What is each device used for?

A lens focuses light to produce an image. A kaleidoscope combines several lenses, resulting in multiple images. Your brain needs accurate images in order to make the body function properly.

1 John 5:19, 20 When you look through a kaleidoscope, you see many images swirling around. It can be very confusing. All those images can make simple things become difficult.

Sometimes the world can be very confusing, surrounding us with many things that draw us away from God. This Scripture reminds us that keeping close to Jesus can help us find God. The more time we spend with him, the clearer everything becomes.

My Science Notes

NAME ______________________________

LESSON 32

ATTRACTIVE IRON

FOCUS Magnetism

OBJECTIVE To explore some properties of matter

OVERVIEW Matter is all around us, but different materials can act very differently. Some materials attract each other. In this activity, we'll explore how this property of matter can be useful.

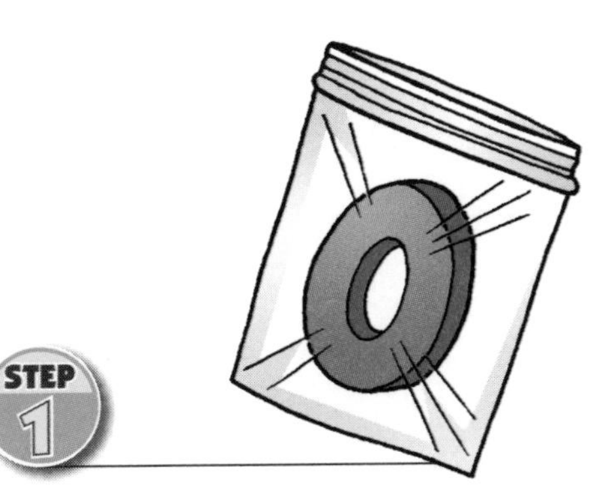

Place the magnet in the plastic bag. **Seal** the bag tightly. (The magnet must stay inside the bag for this entire activity!) **Examine** the bagged magnet. **Record** your observations in your journal.

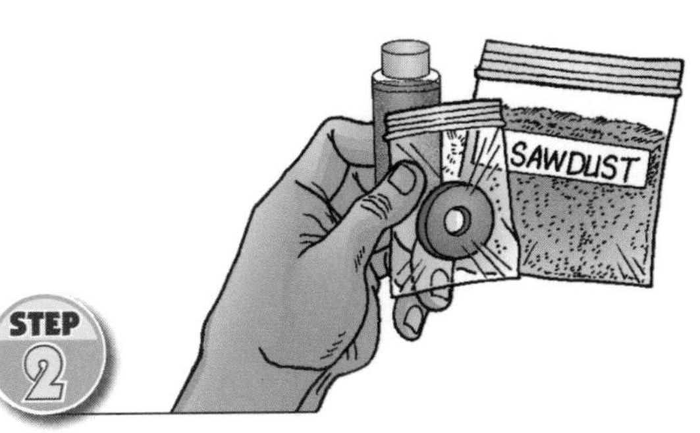

Place the bag of sawdust and the bottle of iron filings on your work surface. **Move** the magnet near the sawdust. **Record** the result. **Move** the magnet near the iron filings. **Record** the result.

Pour the sawdust and iron filings into a non-metal pan. Using a craft stick, **mix** them together thoroughly. **Predict** what might happen if you brought the bagged magnet near this pile.

Move the magnet slowly along the surface of the pile. **Record** the results. Now **drag** the magnet through the pile. **Record** the results. **Replace** the materials in their correct containers. Now **share** and **compare** observations with other research teams.

Some materials are **magnetic**. This means they have the ability to **attract** certain kinds of **metal** like iron, cobalt, and nickel. Scientists call these **ferrous** (magnetic) metals. **Non-ferrous** materials, including some metals like aluminum, are not attracted by magnets. You demonstrated this in Step 4 when you dragged the magnet through the pile. The iron filings were attracted to the magnet, but the sawdust wasn't.

Magnetism is one of the basic **forces** in the universe. Magnets make electric motors run, lock and unlock car doors, operate certain kinds of switches — even check for real coins in a vending machine. Recyclers use huge magnets to help them sort ferrous and non-ferrous materials.

Why was it important to keep the magnet in the bag for this activity? What might have happened if you didn't?

Describe what happened in Step 2. Which material was attracted by the magnet? Why?

What was your prediction in Step 3?
How did this reflect what actually happened in Step 4?

What do scientists call magnetic materials? Is a magnet attracted to all kinds of metal? Why or why not?

Recycling centers receive many different materials, often mixed up in piles. How can a magnet be used to sort these materials? What would you label the two piles?

Conclusion

Different materials behave differently because of different properties. These properties can help scientists and engineers create useful devices. They also help recyclers sort materials.

Food for Thought

Acts 4:31 Until the magnet arrived, the iron filings were lifeless. They didn't show any movement or activity. But when the power of the magnet came near, things changed dramatically! The iron filings really became active, and they helped us see the power that was present.

Scripture tells us that something special happens when God fills our lives. We may feel dull, ordinary, and powerless — but when the Spirit of God begins working in our lives, we can begin to make a real difference in the world. Just as the filings made the magnet's power visible, so we can make God's power and love visible to others.

Journal **My Science Notes**

NAME ______________________

LESSON 33

FERROUS FORCES

FOCUS Electromagnetism

OBJECTIVE To build and use an electromagnet

OVERVIEW In Lesson 32, we learned that magnets attract certain kinds of metals. But did you know there are different kinds of magnets? In this activity, we'll explore a magnet you can turn on and off!

WHAT TO DO

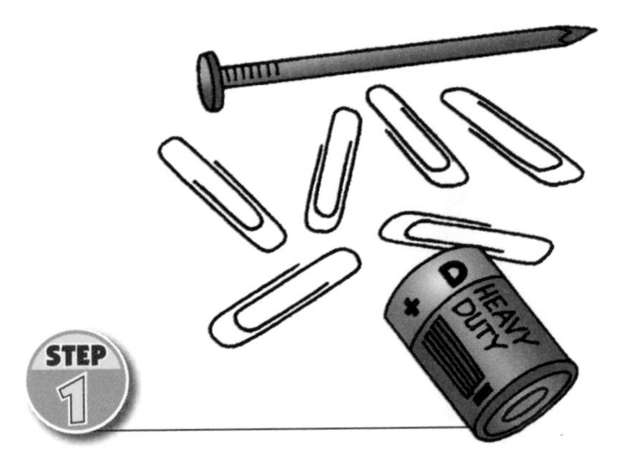

STEP 1

Place the paper clips on your work surface. **Touch** the paper clips with the nail. **Record** the results in your journal. **Touch** the paper clips with the battery. **Record** the results.

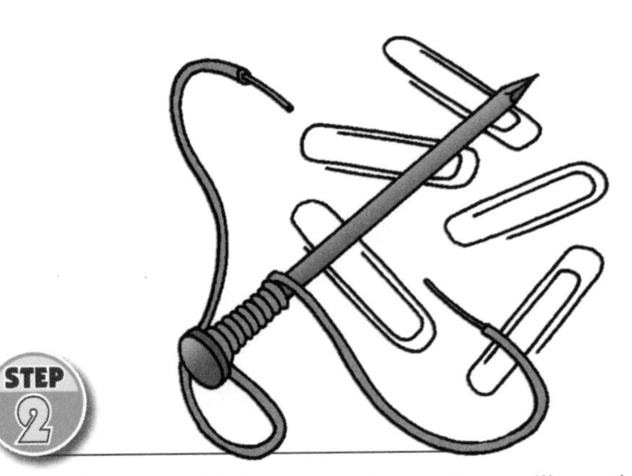

STEP 2

Wrap the wire tightly around the nail (see illustration). **Leave** about six inches of loose wire on each end. **Touch** the tip of the wrapped nail to the paper clips. **Record** the results.

STEP 3

Tape one bare wire end to the positive (+) terminal of the battery, and the other bare wire end to the negative (-) end. **Touch** the nail to the paper clips. **Record** the results.

STEP 4

Remove one wire from the battery, then **repeat** Step 3. **Record** the results. **Review** each step in this activity. **Share** and **compare** observations with your research team.

Ferrous (magnetic) **metals** are not only attracted to a **magnet**, but with the addition of electric **current**, they can be turned into magnets themselves. You created a small **electromagnet** by wrapping wire around a nail (ferrous metal) and passing electric current through it. As long as the current was on, the nail was a magnet!

This activity helped us discover that magnetism has a partner. It's called **electricity**! The magnetism/electricity combination is a basic **force** on Earth. Scientists refer to this as **electromagnetic force**. Electromagnets are used in millions of electric motors, performing many different kinds of work.

Describe what happened in the first two steps. What affect did the battery, nail, and wrapped nail have on the paper clips?

Describe what happened in Step 3.
How was this step different from the previous steps?

3 Compare Step 3 and Step 4. How were they similar? How were they different?

What happens when current is applied to ferrous metal? What does this create?

Based on what you've learned, how could the ability to turn a magnet on and off be helpful?

Conclusion

Magnetism and electricity are related forces. An object made of ferrous metal can be turned into an electromagnet if current is conducted around it.

Luke 24:49 The nail was just a nail until the power of the electric current began to flow. Then it suddenly had the amazing ability to attract other metals to the source of power.

In this Scripture, Jesus tells his followers to wait for the Holy Spirit to flow through them. This gave them the power to draw others to God. When we spend time with God each day, the power of his Holy Spirit can fill our lives. As we become more like Jesus, our example can help draw others to God.

My Science Notes

NAME ______________________________

LESSON 34

MAGIC MATTER

FOCUS Properties of Matter

OBJECTIVE To explore the interaction of matter

OVERVIEW You have five pencils in a cup. You add five more. The cup now holds ten pencils. But does every form of matter "add up" this simply? In this activity, we'll explore another matter combination.

WHAT TO DO

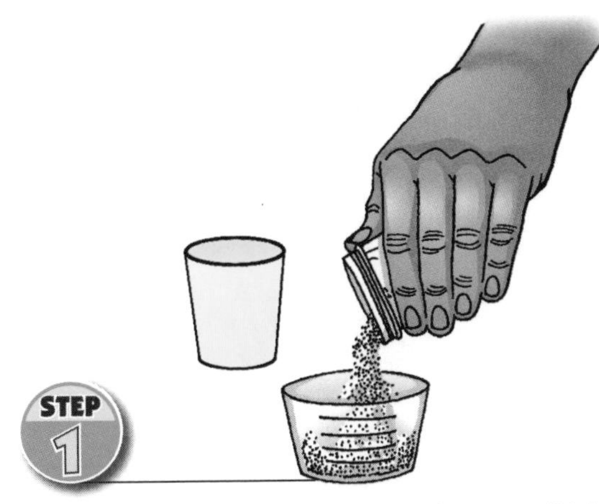

Pour 1/2 cup of salt into a measuring cup. (Make the salt level at exactly 1/2 cup!) **Write** "SALT" in your journal and **record** this measurement. **Pour** the measured salt into a dry paper cup.

STEP 2

Pour 1/2 cup of water into a measuring cup. (Make the water level at exactly 1/2 cup!) **Write** "WATER" in your journal and **record** this measurement.

Predict what the measurement will be if you add the 1/2 cup of salt to the 1/2 cup of water. Will there be exactly one cup, less than one cup, or more than one cup? **Record** your prediction.

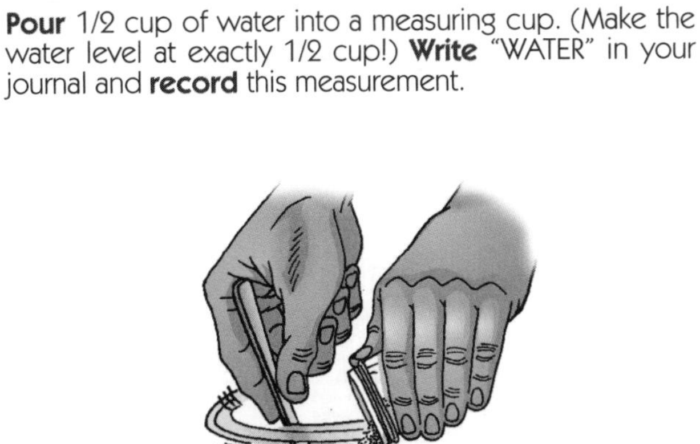

Slowly **pour** the salt into the water while stirring gently with a craft stick. When all the salt is mixed in the water, **check** the measurement and **record**. **Share** and **compare** observations with other research teams.

What Happened?

No, this activity wasn't magic. Like most tricks, it's easy to explain once you understand the science involved. When you poured the salt and **water** together, you made a **mixture**. A mixture is made any time different kinds of **matter** are combined.

As you stirred the mixture, water **dissolved** some of the **salt** breaking it into tiny particles. The water completely surrounded these particles, replacing all the **air** space that had been between the larger particles. This space was part of your measurement in Step 1. Since the air was replaced by water from Step 2, the mixture's final **volume** was less than one cup.

What We Learned

Describe the table salt in Step 1. What state of matter is it?

Describe the water in Step 2. What state of matter is it?

What did you predict in Step 3?
How did this prediction reflect what actually happened in Step 4?

Explain why adding 1/2 cup of salt to 1/2 cup of water didn't result in 1 cup of salt water.

Unlike salt and water, cooking oil and water don't mix. Based on what you've learned, what would the final volume be if you repeated this activity with oil and water?

A mixture is made when two different forms of matter are combined. The volume of a combination can vary depending on the types of matter combined.

Psalm 16:11 Combining the salt and water seemed like simple addition at first glance. You probably didn't think about all the empty space inside matter. The cup didn't fill up quite like you expected!

This Scripture talks about filling up the empty places in our lives. By learning to trust God, we can fill our hearts with joy and peace. Without God, the emptiness results in anger, pain, and sorrow. Which would you like to fill your heart with? The choice is yours!

My Science Notes

NAME ____________________

LESSON 35

FLAME OUT

FOCUS Combustion

OBJECTIVE To explore the "fire triangle"

OVERVIEW Uncontrolled fires can be deadly and destructive. Fire extinguishers can save lives and property. In this activity, we'll explore one way a fire can be extinguished.

WHAT TO DO

STEP 1

Cut a two-liter bottle in half. **Dispose** of the top half as directed by your teacher. **Fill** the dents in the bottom of the bottle half full of baking soda. **Fill** a paper cup with vinegar.

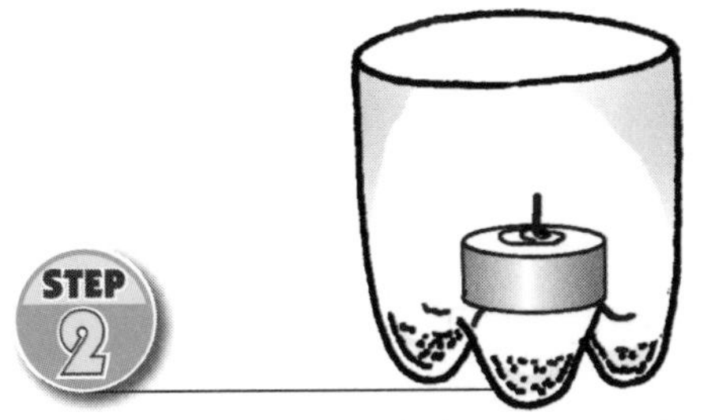

STEP 2

Watch as your teacher lights your candle. **Drip** a little wax onto the center of the bottle. **Blow out** your candle, then quickly **stick** it in the soft wax so that it stands upright.

STEP 3

Make notes in your journal about what you've done so far. Now **watch** as the teacher relights your candle. Gently **pour** the acetic acid from the paper cup onto the baking soda. (Avoid the flame!)

STEP 4

Quietly **observe** the candle. (Don't talk or even move around the container.) **Make notes** about what you see. After a few minutes, **share** and **compare** observations with other research teams.

Fires need three things for **combustion** (burning) to occur — **oxygen**, **fuel**, and **heat**. Firefighters call this the **fire triangle**. Remove one or more parts of the fire triangle and the fire will go out!

This fire started when you supplied heat from a match. The air supplied oxygen, and the candle wax was fuel. In Step 3 you combined acetic acid and baking soda, producing **carbon dioxide** gas. The heavy gas began to fill the container, **pushing** the air with its oxygen out the top. If left undisturbed, this gas soon replaced so much oxygen that the fire went out.

1 **What are the three parts of the fire triangle? What was the source of each part in this activity?**

Describe what happened in Step 3. What did the acetic acid and baking soda combination produce?

Based on what you learned in earlier lessons, which has the greater density: oxygen or carbon dioxide? How do you know this?

Explain why the reaction between the baking soda and vinegar extinguished the fire.

Firefighters sometimes remove brush and trees from an area. What part of the fire triangle does this involve? They also spray water on fires. How does this relate to the fire triangle?

Fires need three things for combustion: oxygen, fuel, and heat. This is known as the fire triangle. Remove any part of the fire triangle, and there won't be a fire.

1 Corinthians 13:13 The fire triangle is a good way to remember what's important when trying to put out a fire! Although each of the parts is different, they are all necessary for combustion to occur.

This Scripture talks about faith, hope, and love. Although each of these is important, Paul reminds us that God's love is the most important of all. When God's love is in our hearts, we think more about the needs of others than our own desires. Reach out to others and share the fire of God's love!

My Science Notes

NAME ______________________________

LESSON 36

PAINTED PETALS

FOCUS Indicators

OBJECTIVE To explore the action of indicators

OVERVIEW Many harmful or helpful things aren't visible. How can you know something is there if you can't see it? In this activity, we'll explore one way to make the invisible visible!

WHAT TO DO

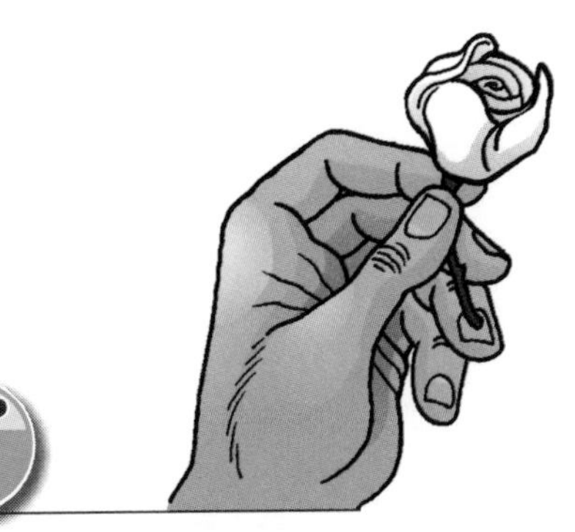

Hold the flower in one hand. **Observe** the flower closely. **Record** your observations in your journal.

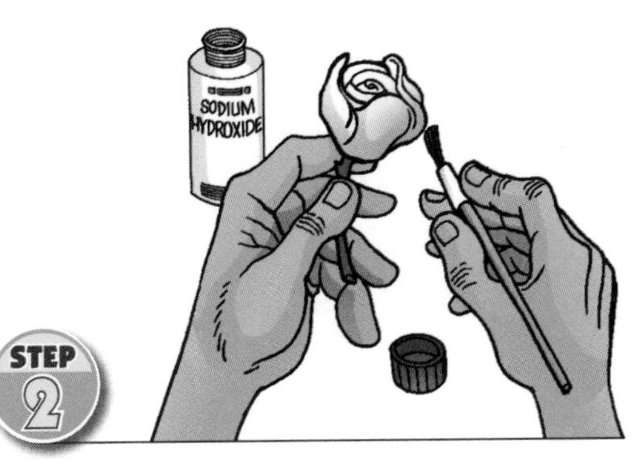

Remove the lid from the sodium hydroxide. **Pour** a little into the lid. **Dip** your paintbrush into the lid and begin to **paint** the flower. **Record** the results.

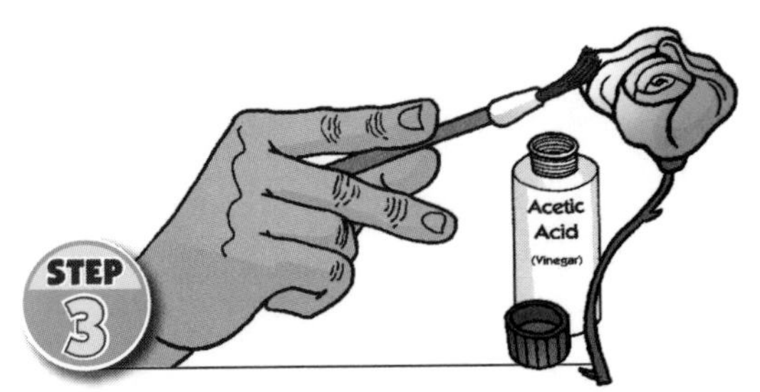

Pour the remaining liquid from the lid into the sink. **Rinse** the lid and paintbrush. **Fasten** the lid on the bottle. **Repeat** Step 2 using the acetic acid bottle. **Record** the results.

Pour the remaining acetic acid from the lid into the sink. **Rinse** the lid and paintbrush. **Fasten** the lid on the bottle. **Share** and **compare** observations with other research teams.

What Happened?

A chemical can be an **acid**, a **base**, or **neutral**. Some chemicals (called **indicators**) change colors around acids or bases. The flower was coated with **phenolphthalein**, an indicator that's colorless around acid. But phenolphthalein turns bright pink in the presence of a base. In Step 2, this helped you see that **sodium hydroxide** is a base. In Step 3, painting on **acetic acid** (vinegar) helped **neutralize** the base, so the indicator became colorless again.

Common acids include vinegar, citrus juice, and soda pop. Common bases include ammonia, detergents, and antacids. Pure water is neutral. Indicators help us check everything from blood sugar levels to swimming pool conditions to soil fertility.

What We Learned

1. **Describe what happened to the flower when you painted it with sodium hydroxide in Step 2.**

2. **Describe what happened to the flower when you painted it with acetic acid in Step 3.**

Most chemicals can be divided into three groups. What are they? Give an example of each.

What is an indicator? How are indicators helpful?

Some berries thrive in acid soil. A farmer buys a field where he wants to plant this kind of berries. Based on what you know, how could an indicator help the farmer?

A chemical can be an acid, a base, or neutral. Indicators help us identify substances by changing colors.

John 1:29 Your flower petals were pure white until the sodium hydroxide came along. Although it looked as clear and harmless as water, it quickly stained the petals pink. But covering the petals with the base removed the stain and made the flowers white again.

Sin is like that. It can look harmless at times, but can really stain your life. Selfish deeds, harmful words, things we do to hurt others — all can leave their mark. This Scripture reminds us that Jesus came to take away the stain of sin. Trust in Jesus and he can give you a clean, new heart.

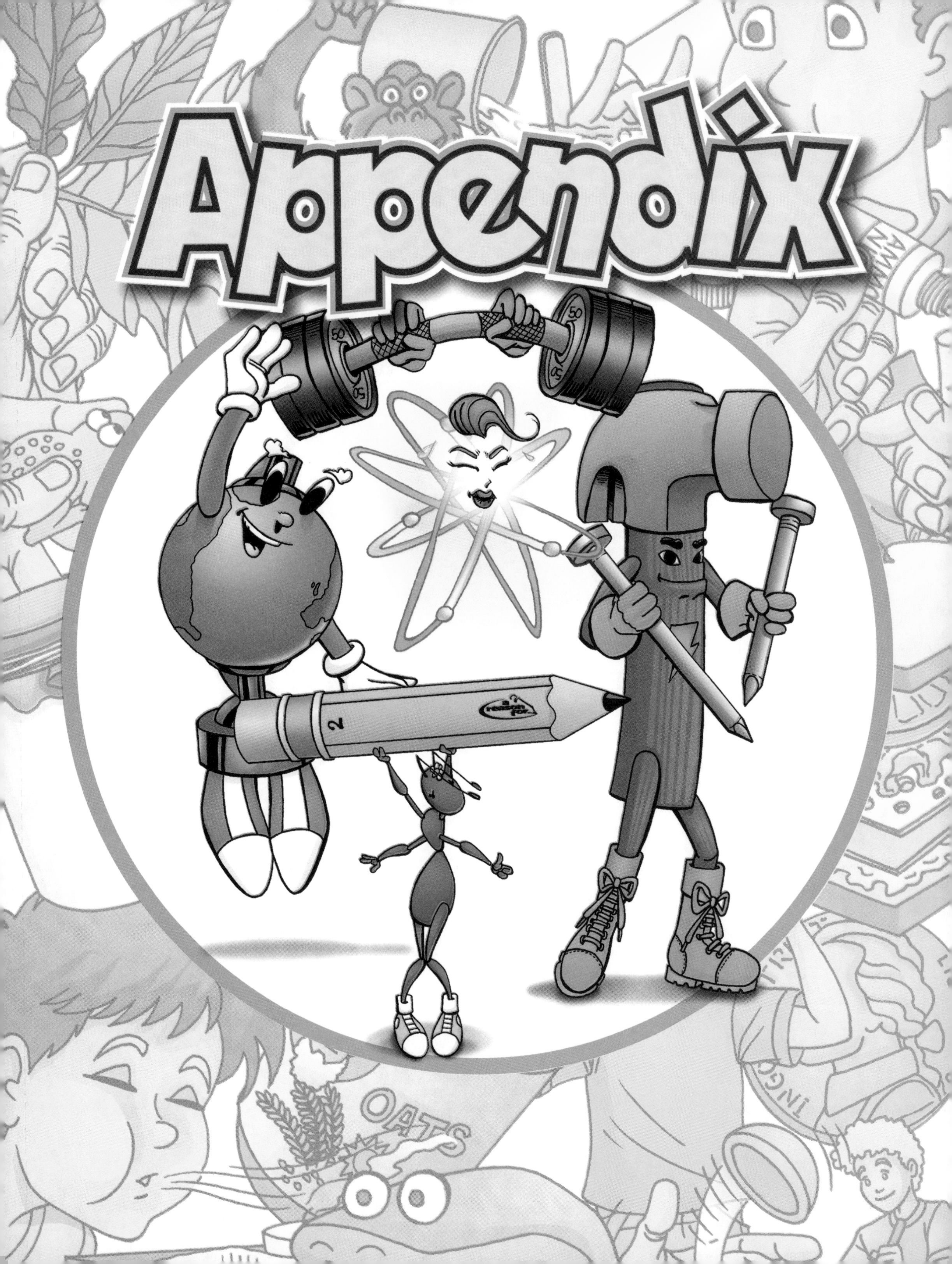
Appendix

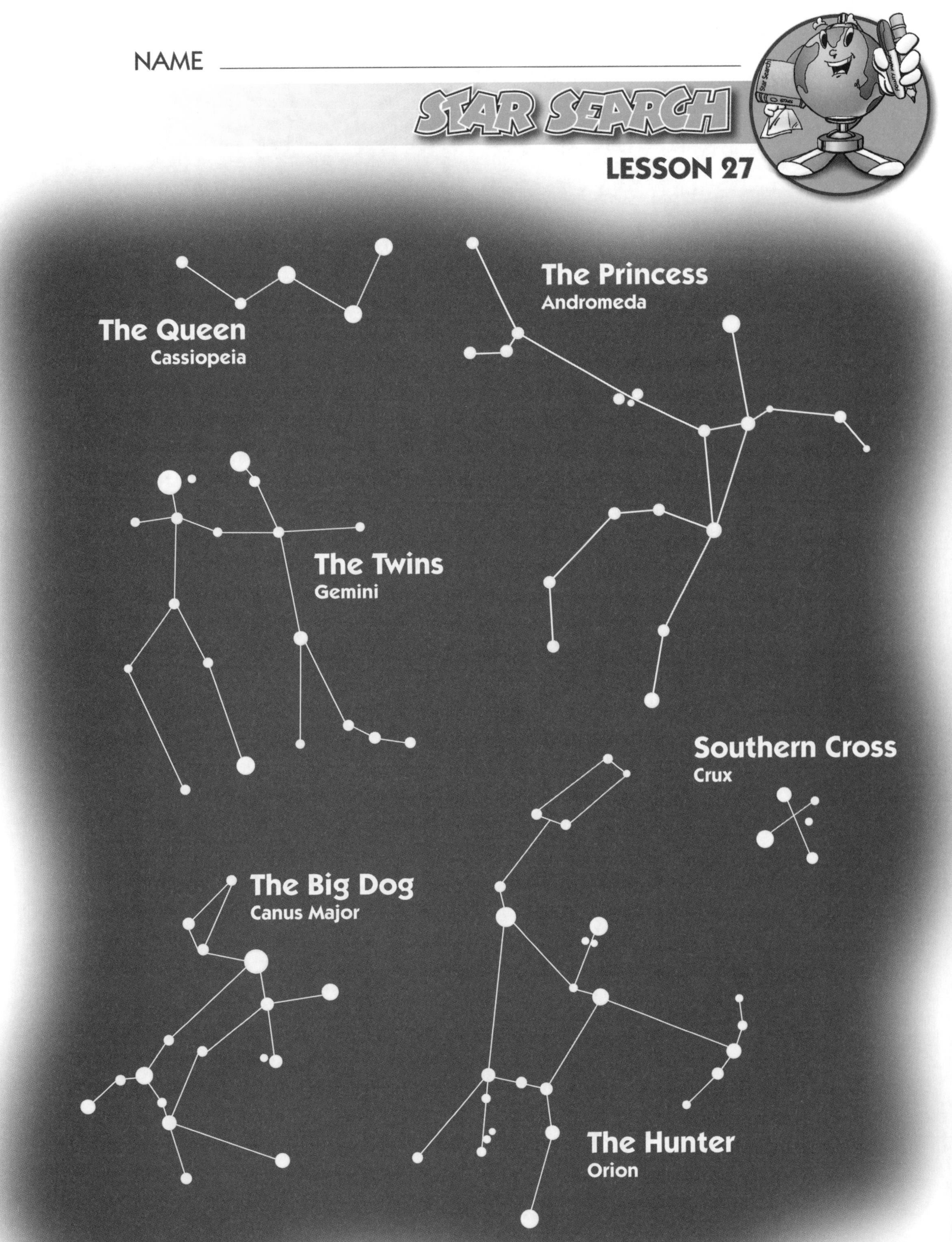
NAME
STAR SEARCH
LESSON 27
The Queen
Cassiopeia
The Princess
Andromeda
The Twins
Gemini
Southern Cross
Crux
The Big Dog
Canus Major
The Hunter
Orion

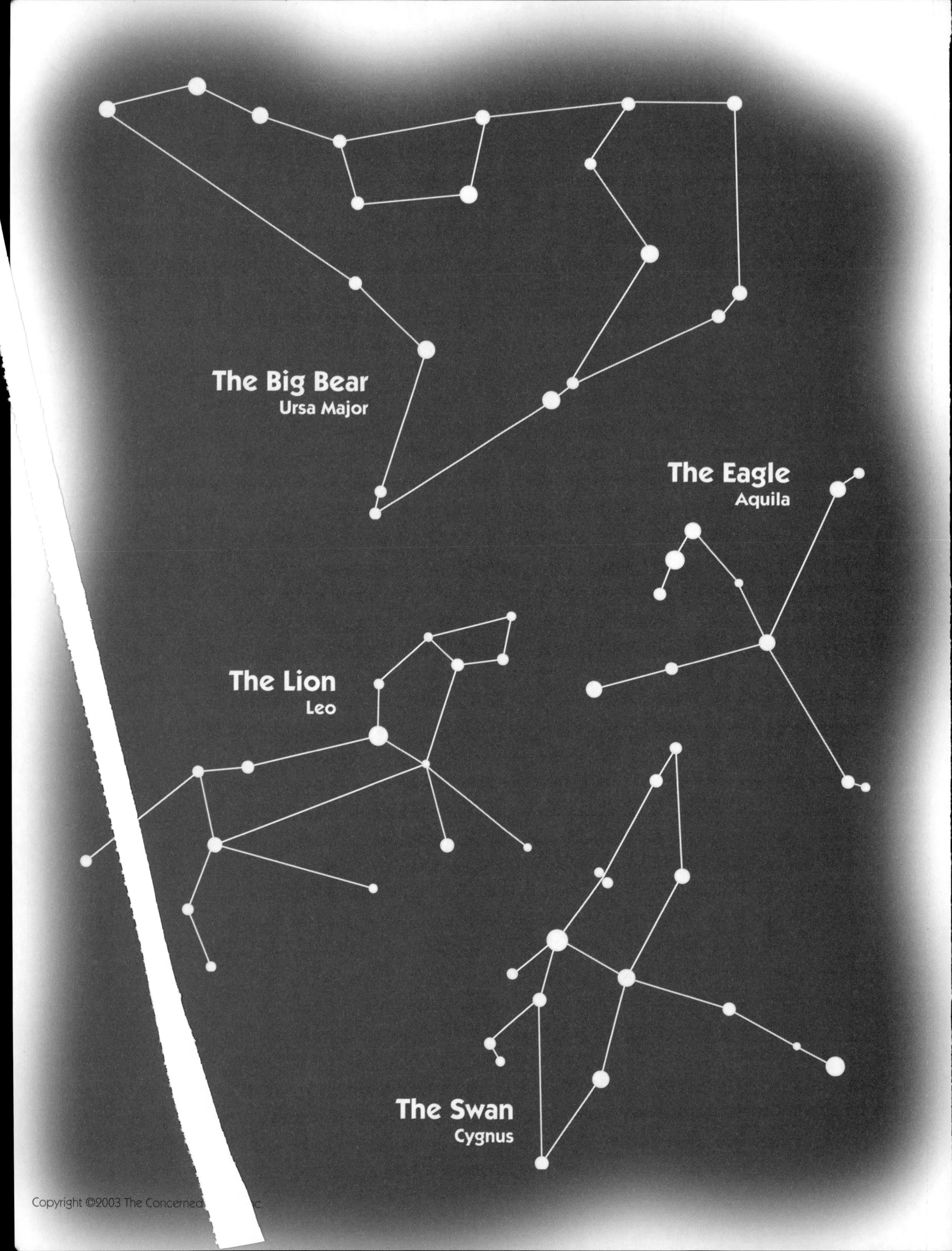
The Big Bear
Ursa Major
The Eagle
Aquila
The Lion
Leo
The Swan
Cygnus

NAME
NOTES

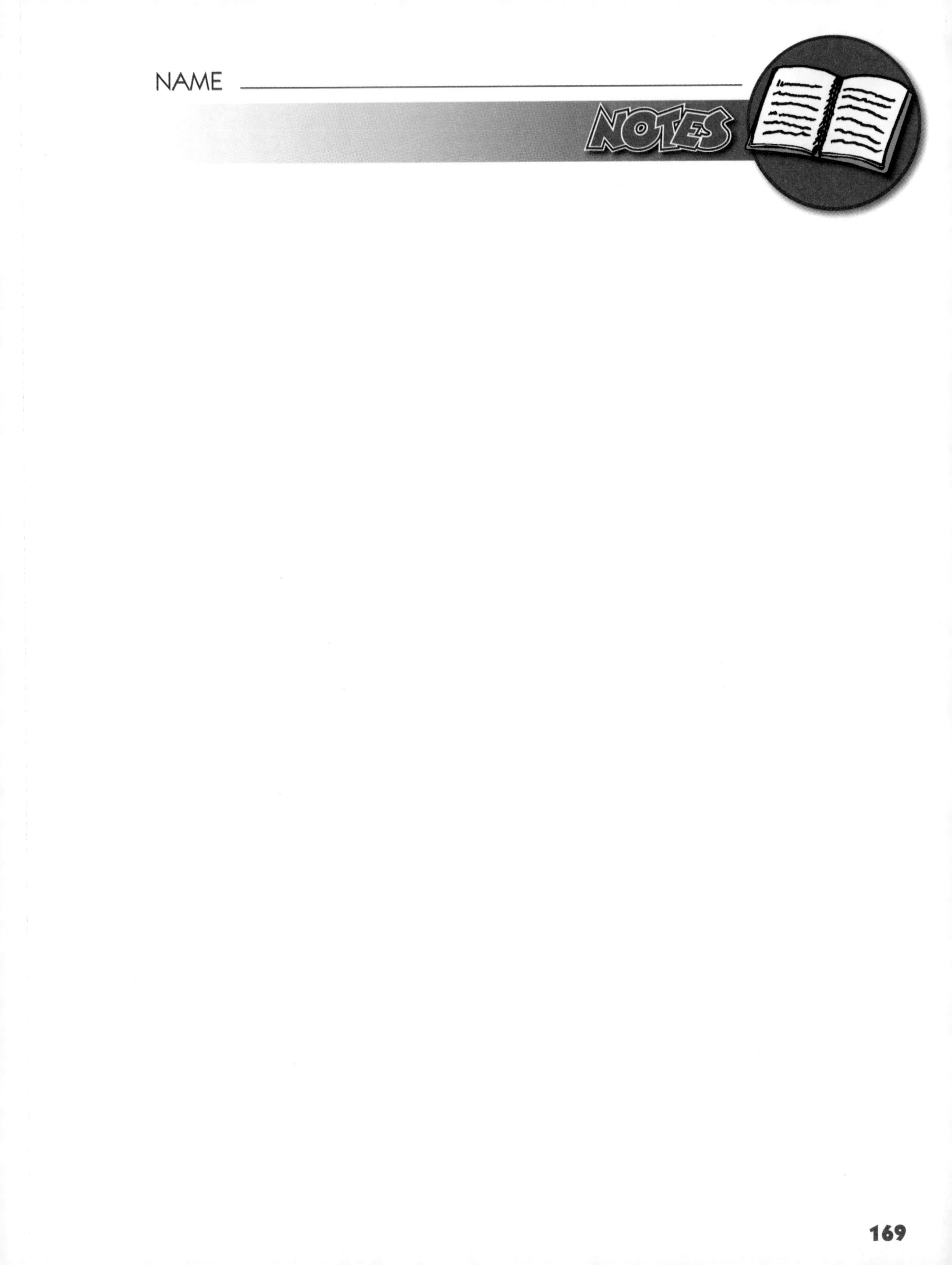
NAME
NOTES

NAME
NOTES

NAME ______________________________